丝路霓裳(中国卷)

Ethnic Costumes Along the Silk Road (China)

/ 韦荣慧 主编

APGTIME 时代出版
时代出版传媒股份有限公司
安 徽 人 民 出 版 社

图书在版编目（CIP）数据

丝路霓裳. 中国卷 ：汉英对照 / 韦荣慧主编. —合肥 ：安徽人民出版社，2020.1

ISBN 978-7-212-10326-2

Ⅰ. ①丝… Ⅱ. ①韦… Ⅲ. ①民族服饰—服饰文化—介绍—中国—汉、英 Ⅳ. ①TS941.742.8

中国版本图书馆CIP数据核字(2018)第290101号

丝路霓裳（中国卷）

Silu Nichang (Zhongguojuan)

韦荣慧　主编

出 版 人：徐　敏　　选题策划：刘　哲　彭新良　陈　娟　　责任印制：董　亮

责任编辑：刘　哲　袁小燕　周　羽　　装帧设计：一方智业

出版发行：时代出版传媒股份有限公司http://www.press-mart.com

安徽人民出版社http://www.ahpeople.com

地　　址：合肥市政务文化新区翡翠路1118号出版传媒广场八楼　邮编：230071

电　　话：0551-63533258　0551-63533292（传真）

制　　版：北京雅昌艺术印刷有限公司

印　　刷：北京雅昌艺术印刷有限公司

开本：889mm×1194mm　1/12　　印张：20　　字数：210千

版次：2020年1月第1版　　2020年1月第1次印刷

ISBN 978-7-212-10326-2　　定价：666.00元

序言

2100多年前，张骞出使西域，开启了中国同中亚各国友好交往的大门，开辟出一条横贯东西、连接欧亚之路。19世纪70年代，德国地理学家李希霍芬将之命名为“丝绸之路”后，即被广泛接受。古代中国是最早种桑、养蚕、生产丝织品的国家，而公元前5世纪希腊艺术中一些雕塑女神所穿衣服都是柔软精细的丝质面料，西方学者由此考证在公元前五六世纪，中国丝绸已经传至地中海沿岸国家。同时，丝绸之路也为中国带来异域文化，或多或少影响中国服饰的审美与风格。除此之外，丝绸之路在商品交流、文化交流、技术交流等过程中对中国也产生了较大的影响，为世界服装的发展，为促进不同国家、不同民族、不同文化相互交流与合作做出了重要贡献。千百年来，在这条古老的丝绸之路上，各国人民共同谱写出千古传诵的友好篇章。

传统的丝绸之路，起自中国古代都城长安，途经甘肃、新疆，再经中亚国家、阿富汗、伊朗、伊拉克、叙利亚等而达地中海，以罗马为终点。这条线路全长6440公里，因为要途经新疆、中亚地区广阔的沙漠、戈壁，因此也被称为“沙漠丝绸之路”。“沙漠丝绸之路”在世界史上有重大的意义，它不仅是亚欧大陆的交通动脉，也是中国、印度、希腊三种文明交汇的桥梁。回族、维吾尔族、哈萨克族、塔塔尔族、塔吉克族、乌孜别克族、柯尔克孜族、俄罗斯族、锡伯族、裕固族、东乡族、撒拉族、保安族等少数民族长期生活在“沙漠丝绸之路”沿线。

草原丝绸之路是指蒙古草原地带沟通欧亚大陆的商贸大通道，是丝绸之路的重要组成部分。草原丝绸之路的主要路线，是由中原地区向北越过古阴山（今大青山）、燕山一带长城沿线，往西北方向穿越蒙古高原、南俄草原，直至中亚、西亚、地中海沿岸。草原丝路的形成，与自然生态环境有着密切的关系。在整个欧亚大陆的地理环境中，只有在北纬40度至50度之间的中纬度地区，才有利于人类的东西向交通，而这个地区就是草原丝绸之路的所在地。这里是游牧文化与农耕文化交汇的核心地区，是草原丝绸之路的重要连接点。草原丝路在历史上扮演着重要的角色，其形成、发展和繁荣代表了中国历史的一个辉煌时期。作为中西文化交流的产物，草原丝路一直被视为对外交流的经典，对研究中西经济、文化发展起到了重要作用。从文化传播的角度看，草原丝路文化传播是全方位的，所经过的地区是人类生活的聚集区，文化的冲击力与波及面较大，而游牧民族迁徙的特点使得文化的传播速度较快，持续性更长久。草原丝路沿线的各民族服饰，主要包括蒙古族、满族、达斡尔族、鄂温克族、鄂伦春族等少数民族服饰。

海上丝绸之路，是古代中国与外国交通贸易和文化交往的海上通道，也称“海上陶瓷之路”和“海上香料之路”。海上丝绸之路形成于汉武帝时期。隋唐以前，它只是陆上丝绸之路的一种补充形式；到了唐代，由于西域战火不断，陆上丝绸之路被战争阻断，代之而兴的便是海上丝绸之路。宋代以后，随着我国造船、航海技术的发展和经济重心的南移，从广州、泉州、杭州等地出发的海上航路日益发达，越走越远，从南洋到阿拉伯海，甚至远达非洲东海岸，人们把这些海上贸易往来的各条航线，通称为“海上丝绸之路”。海上丝路是已知最为古老的海上航线，分为东海航线和南海航线两条线路。南海航线，又称“南海丝绸之路”，起点主要是广州和泉州，途经东海、南海、印度洋，直至日本、朝鲜、东南亚、南亚、东非。这条途经南海的航线是海上丝路的主线。东海航线，也叫“东方海上丝路”，由胶东半岛出发，直通辽东半岛、朝鲜半岛、日本列岛直至东南亚，它在海上丝路中居次要的地位。居住在海上丝路沿线的有黎族、畲族、台湾少数民族等。

服饰是人类特有的劳动成果，服饰文化是人类文化遗产的重要组成部分。人类社会经过蒙昧、野蛮到文明时代，缓缓地行进了几十万年。我们的祖先在与猿猴相揖别以后，披着兽皮与树叶，在风雨中徘徊了难以计数的岁月，终于艰难地跨进了文明时代的门槛，懂得了遮身暖体，创造出一个物质文明。几乎是从服饰起源的那天起，人们就已将其生活习俗、审美情趣、色彩爱好，以及种种文化心态、宗教观念，都沉淀于服饰之中，构筑成了服饰文化的精神文明内涵。

服饰文化具有的深厚内涵，蕴含着视觉冲击力和情绪感染力。服饰文化可以超越国界为世界人民所接受，容易引起不同国家、不同种族人群的喜爱和共鸣，容易拉近中国人民与世界人民的距离。它可以突破社会制度和意识形态造成的鸿沟，达到从文化亲近走向政治亲近和经济合作。

随着全球化时代的到来和中国同世界各国关系的快速发展，古老的丝绸之路重新焕发出生机与活力。2013 年，中国国家主席习近平提出了“一带一路”重大倡议，承古惠今、连接中外，赋予古老的丝绸之路新时代内涵。作为丝绸之路起点的国家，中国是一个统一的多民族国家，共有 56 个民族。2000 多年来，各民族休戚与共，唇齿相依，秉承开放包容的姿态以及天人合一的哲学观，传承与弘扬着古老的丝路精神。特别是 56 个民族灿烂夺目的服饰及其背后的文化，在世界上独树一帜，为人类留下一笔珍贵的文化遗存。相信本书展现的绚丽的中国民族服饰，将再一次吸引海外文化界、艺术界和时尚人士的目光，为新时代的“一带一路”增光添彩，唤起我们对丝绸之路上许多传奇故事的回忆，让古老的丝路文化、多彩的民族风情与色彩斑斓的霓裳，成为“一带一路”上永远的风景线。

Preface

More than 2,100 years ago, Chinese diplomat and envoy Zhang Qian of the Han Dynasty helped establish the Silk Road, a network of trade routes that linked China to Central Asia and Europe. This Silk Road opened windows of friendly engagement among nations, adding a splendid chapter to the history of human progress. Because silk comprised a large proportion of the trade along this ancient road, in 1877, it was named as the "Silk Road" by Ferdinand Freiherr von Richthofen, an eminent German geographer.

"The Maritime Silk Road", like its overland counterpart, played an important role in trade and communication as well as culture. This maritime silk road had its origins during the Qin and Han Dynasties, developed in the period of the Three Kingdoms and the Sui Dynasty and flourished in the Tang and Song Dynasties until the Ming and Qing Dynasties. It was the most well-known ancient marine navigation route. It was the lifeline for fishermen, commercial line for businessmen and pilgrimage line for religious believers.

In 2013, President Xi Jinping proposed the initiative of the Belt and Road. This great strategical proposal revitalizes the spirit of peace and cooperation of the ancient silk routes and benefits the economic cooperation, including policy coordination, trade and financing collaboration between nations along the routes and over the world.

These ancient trade routes were named as Silk Road by German geographer Professor Ferdinand Freiherr von Richthofen, because the main and important commodities of the trade were the bulk of raw and processed silk. According to archaeological study, the ancient statues of Greek goddesses worn silk and silk was carried from China as early as in the 5th or 6th century BC. China was the only country that produced silk at that time. Silk from China was not only the material for costumes, but also a showcase of China's civilization and Chinese culture of garments. China's silk had contributed greatly to the development of world costumes.

The ancient silk routes were not for trade only, they boosted flow of knowledge as well. Chinese costume design and patterns were influenced by those from foreign countries. For instance, the Persian patterns can be seen on the murals in the Dunhuang Grottoes. The well-known costumes of the Tang Dynasty were colorful, rich in design and various in style, because they were inclusive of those patterns and designs from India, Persia and those of central and north Asian countries.

Besides, these engagements had a great influence on China's costumes and ornaments. Through trade, fur, feather, and jewelry found their way to China. In history, most agate was from central Asia, India and Persia. The technical exchange of weaving and dyeing was helpful for China's silk production, based on its traditional producing experience, to become a full range production both in breeding and processing as well as techniques and varieties of products.

Costumes are so important to mankind that they stand as a proof of transition from barbarism to

古代丝绸之路示意图（来源：中国一带一路网）
The Silk Road

civilization. They are part of human cultural heritage. In addition to their practical value of protecting human body, they embody the pursuit of beauty of humankind. The cultural connotation of living habits, love of colors, and religious concepts can be seen and traced in costumes.

With profound cultural connotation, visual impressiveness and attractive nature, costume cultures can communicate without language barriers in international cultural exchange. Exhibitions on costume culture have been on display and are highly appreciated around the world.

The speedy economic development and the opening-up policy give a better opportunity to revitalize the costume culture that has a long and splendid history. The ethnic costumes of China are special and unique in the world for the colorfulness, diverseness and profound cultural connotation. We expect that this album will attract the attention of the circles of culture, art and fashion. The reminiscences of those historical chapters on the Silk Road and the glimpse of colorful costumes and folk arts along the Silk Road are always the beautiful scenery along the Belt and Road.

目录 Contents

（按中华人民共和国国家民族事务委员会2010年人口统计数排序）

汉族
Han

汉族服饰，与中华文明一脉相承，阶段明显，包容并蓄，善于融合创新。丝绸及云锦、宋锦、蜀锦都深远地影响了中国乃至世界的服饰文化，至今依然华彩四射。

中国自古以来是一个统一的多民族国家，共有56个民族。汉族是中国的主体民族，历史悠久，人口众多。据2010年的人口统计，有12亿多人，约占世界总人口的18%，占中国总人口的92%，也是世界上人口最多的民族。

汉族的服饰文化历史源远流长。经历5000多年各个朝代的发展变迁，创建了一套体现其礼仪风俗、审美品格、哲学思想等文化的服饰体系，为人类的生活方式及服饰文明发展做出了巨大贡献。

21世纪初，经过40多年改革开放，中国综合国力显著增强，文化自信提升，人们开始积极审视传统汉族服饰文化中的优秀特色，创新设计在中国蔚然成风。新时代中国服饰文化发展生机勃勃，呈现出现代化、融合化的特点。越来越多的新创汉式服装融入人们的日常生活中。

China has been a united multi-ethnic country since ancient times. There are totally 56 ethnic groups and the Han ethnic group has the largest population in China as well as in the world. According to the 2010 national population census, the total population of the Han people is over 1.2 billion that accounts for more than 92% of China's population, about 18% of the total of the world. The Han people are descendants of the ancient Huaxia people and other peoples in China. The Han people has a long history without interruption and it is the principal ethnic group in China.

With a history of more than 5000 years, traditional Chinese clothes reflect the development from barbarism to civilization and embody clothing etiquette, aesthetic standard, ideology and philosophy. The traditional Chinese clothes vary according to different dynasties and locales. Although many dynasties feature their own clothes, some of which were really exquisite and beautiful, the clothing, hat wearing and dress code in the Han Dynasty (206 BC—220 AD) are always considered as the characteristic clothing for the Han ethnic group, hence the name of Hanfu is popularized recently and refers to the historical dress of the Han people.

A long flowing robe with loose sleeves and a belt at the waist is the typical dress of Hanfu. In history, the dress style varied in different dynasties but not in a radical way until the Qing Dynasty (1616—1911) when the Manchus started ruling China. The Han people were forced to change their hair style and clothing style. The Han people were required to wear Manchus' long gown, cheongsam, and mandarin jacket. With this restriction loosened soon afterward, traditional Han clothes, Manchu clothes and part of Western-style clothes coexisted. After the overthrow of the Qing Dynasty, the Han people wore western-style attires and those clothes integrating traditional Chinese and western style like Zhongshan suit (or Mao suit).

Now more and more people like to wear the traditional-style garments during festivals and important occasions to express their pride in the traditional culture and show their support to revitalize the Han clothing.

汉代《人物夔凤图》（局部）
"Dragon, phoenix and a human figure", a painting of the Han Dynasty

唐代周昉绘制的《簪花仕女图》体现出唐代衣冠的华丽神韵。

"Women with flower hairpins", a painting of the Tang Dynasty by Zhou Fang, portraying the resplendent costumes

北宋赵佶绘制的《听琴图》，其人物服饰典雅大方。
"Listening to music", a painting by Zhao Ji, the emperor of the Northern Song Dynasty. The costumes are graceful and elegant.

明代唐寅绘制的《王宫蜀妓图》，其人物服饰延续了宋代服饰的典雅精致。

"Maids in Palace", a painting of the Ming Dynasty by Tang Yin. The costumes continued the gracefulness and elegance of the Song Dynasty.

明代王绎绘制的《杨竹西小像》中人物衣着是明代汉族服饰的基本款式。
A painting by Wang Yi of the Ming Dynasty, the portrayal depicting the basic design of costumes then.

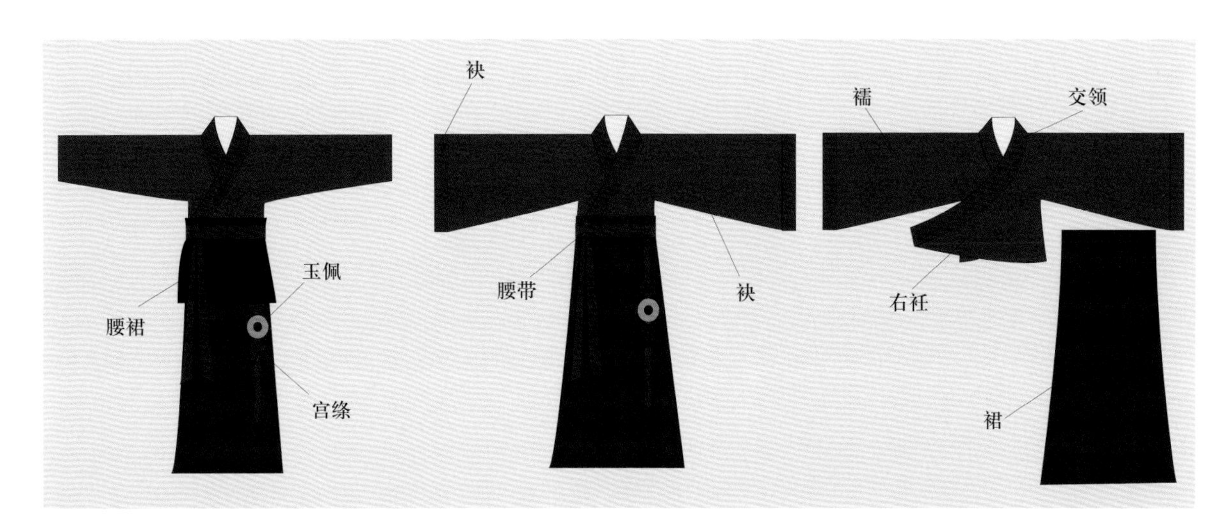

汉装明显有别于其他民族服饰的关键款式为：交领右衽（即衣领呈 y 形），宽袍大袖，系带隐扣。
The basic characters of the traditional Han costumes: cross collar with right lapel ("y" shaped collar), loose gown and sleeves, hidden lacing.

汉族服饰的优秀传统在新时代有了自信的认识和精彩的创意发挥，民族性、艺术性、品质感皆融入现代生活服饰和现代时尚设计中。

With the confident understanding and creative development of the tradition, the modern garments integrate tradition and fashion.

丝绸与织锦

丝绸就是蚕丝织造的纺织品。纯桑蚕丝所织造的丝绸，又被特别称为“真丝”。

丝绸起源于中国，是中国的特产。中国人发明并大规模生产丝绸制品，更开启了世界历史上第一次东西方大规模的商贸交流，史称“丝绸之路”。中国也因此被称为“丝国”。

汉民族丝织提花技术起源久远，早在殷商时代就有丝织物。周代丝织物中开始出现织锦，花纹五色灿烂，技艺臻于成熟。其中云锦、蜀锦和宋锦成为汉民族最具代表性的丝织工艺。

成都蜀锦、南京云锦、苏州宋锦从及广西的壮锦，并称中国的四大名锦。2009 年南京云锦被列入联合国教科文组织《人类非物质文化遗产代表作名录》。

Silk originated from China. Legend has it that in ancient times, the wife of Emperor Huang Di taught people how to raise silkworms and how to extract the silk. The silk trade started as the first trade route that linked the east and west, known as the “Silk Road”. With the development of silk production, the pattern design, weaving, embroidery and dyeing skills have been improved, and thus many silk products of beauty and artistry like brocade become the well-known silk arts. Brocades made in Chengdu, Nanjing, Suzhou and Guangxi are acclaimed as the best four brands of brocade in China. The Nanjing brocade has been included in the catalogue of UNESCO’s World Heritage.

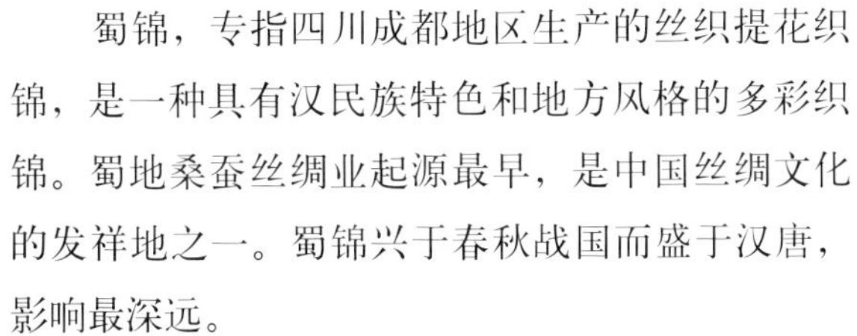

蜀锦，专指四川成都地区生产的丝织提花织锦，是一种具有汉民族特色和地方风格的多彩织锦。蜀地桑蚕丝绸业起源最早，是中国丝绸文化的发祥地之一。蜀锦兴于春秋战国而盛于汉唐，影响最深远。

Shujin is the brocade produced in Chengdu, Sichuan Province. The area in Southwest China has a long history in producing silk. The brocade there has a style of both traditional and local characteristics.

宋锦，主产于苏州，故称“苏州宋锦”。苏州宋锦织工精细，色泽华丽，图案精致，质地坚柔，艺术格调高雅，具有宋代以来汉民族的传统风格与特色。

Songjin is the brocade produced in Suzhou. The style of exquisite workmanship, fine design and colorful patterns well inherited those characters from the Song Dynasty.

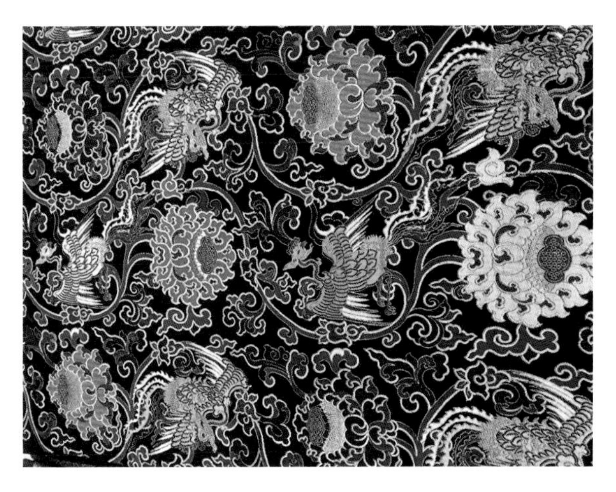

云锦，主产于南京，因其色泽绚烂，美如云霞而得名。南京云锦有“寸锦寸金”之誉，用料考究，织造精细，图案精美，格调高雅，达到了丝织工艺的巅峰状态，代表了中国丝织工艺的最高成就，是中国丝绸文化的璀璨结晶。

Yunjin is the brocade produced in Nanjing. This brocade is well-known for its design, workmanship and quality. It represents the best of silk brocade and silk production of China.

南京云锦传统丝织工艺传承至今。
The brocade producing technique and workmanship of Nanjing brocade has been well inherited.

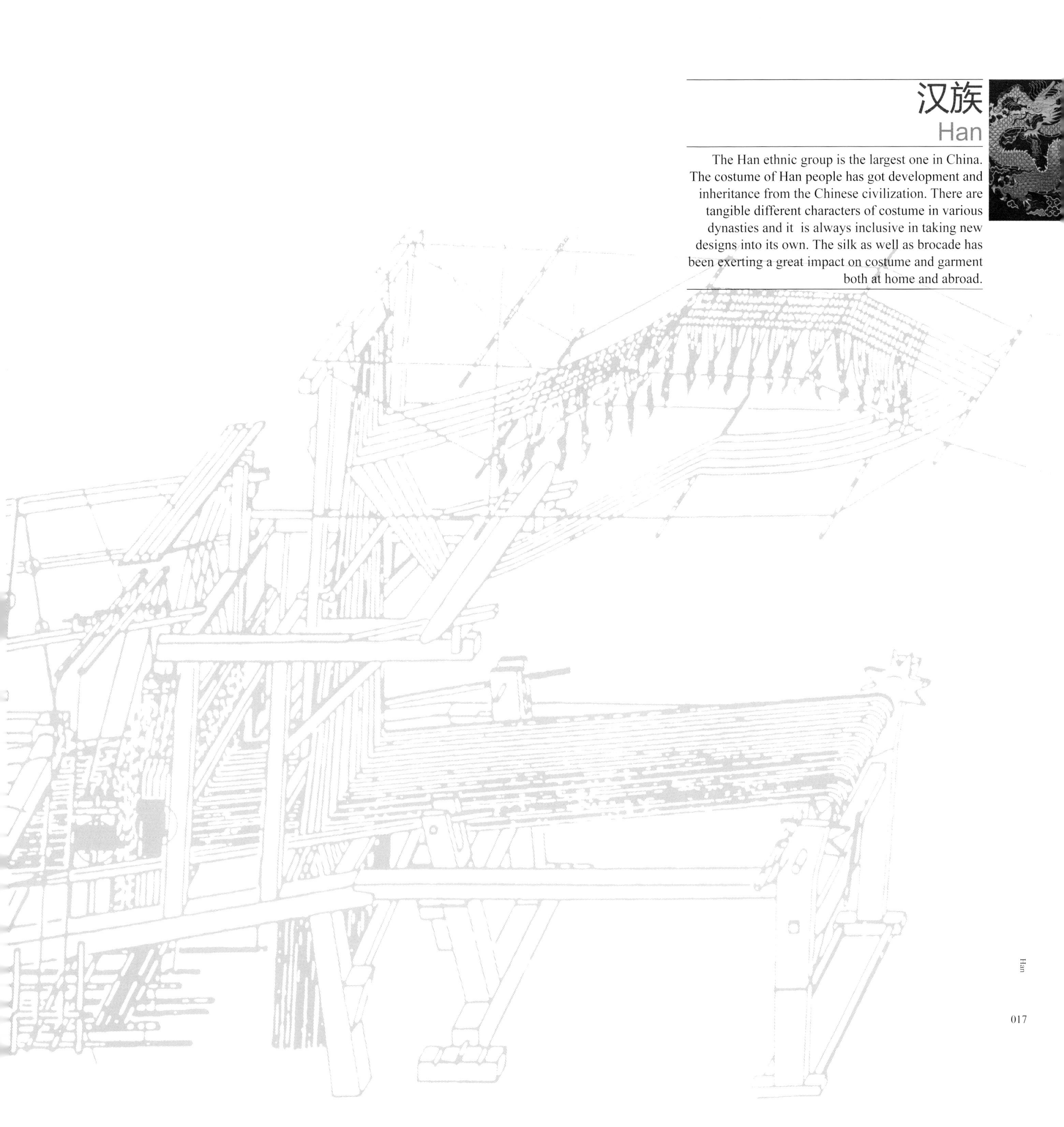

汉族
Han

The Han ethnic group is the largest one in China. The costume of Han people has got development and inheritance from the Chinese civilization. There are tangible different characters of costume in various dynasties and it is always inclusive in taking new designs into its own. The silk as well as brocade has been exerting a great impact on costume and garment both at home and abroad.

壮族
Zhuang

壮族妇女擅长纺织和刺绣。壮锦被誉为中国四大名锦之一。壮锦源于宋代，是中华民族的文化瑰宝，壮文称之为天纹之贡。

壮族是中国少数民族中人口最多的一个民族，主要聚居在中国的广西壮族自治区，人口共有1692.6万人。壮族有本民族的语言壮语，也有本民族文字壮文。

壮族男子服饰比较素，女子服饰则多姿多彩。壮族传统服饰中头巾多由土布或壮锦制成。壮族人民将自己的生活情景融入织物中，色彩艳丽，构思巧妙，既有飞禽走兽、花草虫鱼，也有山水名胜以及人物，富含特有的壮族文化。

壮锦是壮族服饰重要的组成部分。

The Zhuang ethnic minority is China's largest minority group. Its population is over 16 million. Most of the Zhuangs live in the Guangxi Zhuang Autonomous Region. Zhuang brocade is a splendid handicraft. The designs include beautiful scenery and exquisite patterns. In traditional style, Zhuang men wear coats or jackets with two large-size pockets in the front. The trousers are loose and long. Women like collarless, embroidered and trimmed jackets buttoned on the left together with baggy trousers, embroidered belts and shoes and pleated skirts.

黑衣壮
Black-costume Zhuang

壮族
Zhuang

Zhuang brocade, starting in the Song Dynasty, is well-known as one of the four famous brocades in China.

色彩艳丽的壮锦是承载着壮族人记忆的"活化石"
Zhuang brocade, the colorful brocade considered as the "living fossil" of this ethnic group's history

回族

Hui

回族服饰是回族生存环境、文化活动的生动写照，也是回族文化传承的重要载体。

回族是中国人口较多的一个少数民族，人口共有1058.6万。宁夏回族自治区是其主要聚居区。伴着漫漫丝绸之路上的驼铃声和大海中商贸船舶的桅杆风帆，回族先民跨入中华民族序列。

回族生活区域遍布中国，由于长期和汉族杂居，通用汉语，不同地区持不同方言。

回族通常称服饰为“衣着”“穿戴”。由于地域、性别的差异，形成了回族服饰的不同风貌和特色，其中最有特色的是男子戴的白色帽子和女性常戴的盖头。地处贵州威宁与云南会泽、宣威及毗邻地区的回族，男子大多包青、蓝、白布的大套头，而云南昭通地区的回族，男性包白帕而不露头顶，或戴白帽不折边，阿訇戴的帽子帽顶上有经文。少女不包包头而戴绣花“勒子”或顶红、绿方巾。

With over 10 million population, the Hui ethnic minority is the most widely scattered ethnic minority in China. Their costumes and dressing habits are influenced by their religious belief and the local customs. Generally, they like white because it symbolizes holiness and purity.

回族
Hui

Hui costume reflects the culture and living environment of this ethnic group.

满族
Manchu

满族服饰有很强的民族传统特色，但满族长期与汉族杂居，在服装款式、服饰色彩与服饰图案上都有不同程度的演变，特别是与汉文化及服饰的融合，产生了现代旗袍、新唐装等。

满族主要生活在中国东北地区，人口共有1038.7万人。满族有自己的语言和文字。

满族妇女有“辫发盘髻”的习俗，也是来自女真人留下的传统。大拉翅是清宫妇女装扮使用的一种头饰，形似扇面，内用铁丝或藤条按照头围的大小制作成圆箍固定，外包青缎或青绒布。它能够将翠扁方、金钗、凤簪、珍珠流苏等首饰同时插戴。

满族的祖先有“削木为履”的习俗，最著名的则为旗鞋，在鞋底的中间做跟高5～10厘米的木底，鞋面为刺绣的绸缎，又称“马蹄鞋”“花盆鞋”。妇女盛事时多穿着花盆鞋，装饰旗头，走起路来婀娜娴雅。

With a population of over 10 million, the Manchu ethnic group is the second largest minority and scattered wide in China. The traditional costumes of male Manchus are a narrow-cuffed short jacket over a robe with a belt at the waist to facilitate horse-riding and hunting. The Cheongsam, known as “qipao”, a form-fitting Chinese dress for women, is derived from the Manchu women’s long gown.

辽宁省欢度满族民间节日

Manchu people in Liaoning Province are celebrating their traditional festival.

满族刺绣
Embroidery of Manchu

满族
Manchu

Traditional Manchu costume has absorbed those of Han attire and evolved into contemporary style, such as "qipao", cheongsam, and "Tangzhuang", Tang suit.

满族马褂
Manchu jacket, or Mandarin jacket

维吾尔族

Uygur

维吾尔族喜欢戴绣花帽，着绣花衣，穿绣花鞋，扎绣花巾，背绣花袋，衣着服饰色彩斑斓。

维吾尔族主要聚居在中国的新疆维吾尔自治区，人口共有 1006.9 万人。维吾尔族有自己的语言和文字。

受丝绸之路的影响，其传统服饰至今仍然保留着丝绸面料的特点。维吾尔族男女外出时喜爱戴小花帽、绣花帽，维吾尔族称之为“多帕”。它不仅是维吾尔族群众日常生活的一种特色服饰，也是一种精美的工艺品。花帽主要有巴旦姆花帽、塔什干花帽、格来木花帽、奇曼花帽、曼波尔花帽、阿克多帕等种类。

在纺织品中，维吾尔族妇女最喜欢的衣料是“艾得丽丝绸”。这是一种富有浪漫色彩的丝绸，它采用古老的扎染法，按图案的要求，在经线上扎结进行染色。

The population of Uygur is over 10 million and most of them inhabit in the Xinjiang Uygur Autonomous Region. The Uygurs' cotton growing and cotton yarn spinning industry has a long history. They usually wear cotton garments. Men wear a long gown called “qiapan” which opens on the right and has a slanted collar. It is buttonless and has a long square cloth band bounded around the waist. Women wear broad-sleeved dresses and black-waist coats with buttons sewn in the front. The Uygur, old and young, men and women, like to wear a small cap with four pointed corners, embroidered with black and white or colored silk threads.

维吾尔族艾得丽丝绸中大面积运用绿色
Colorful Uygur silk

维吾尔族男子服饰
Uygur men's attire

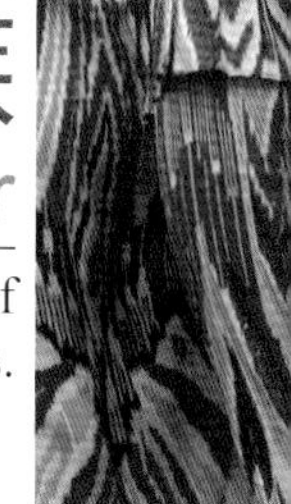

维吾尔族
Uygur

Uygur costume is colorful. People are fond of embroideries.

和田维吾尔族妇女的服饰喜用黑色、蓝色和白色。
Uygur women in Hetian area like costumes in black, blue and white.

苗族
Miao

苗族传统服饰被誉为“穿在身上的史书”，它记录了上古文明的原貌和本民族迁徙发展的历程，也创造了苗族独特的审美生活。刺绣、银饰和蜡染是苗族古老、精湛的服饰工艺。

苗族主要分布在中国贵州、湖南、云南、广西等省区，人口共有942.6万人。苗族有自己的语言苗语，历史上没有本民族文字，现在创有苗文。

苗族传统服饰被誉为“穿在身上的史书”。男性穿无扣短衣或长衫，着青色长裤、包头。刺绣和银饰是苗族既古老又精湛的工艺，有盛大的节日或是婚嫁时，女性将银饰钉在绣着古老传说的传统服装上，苗族姑娘胸前佩戴硕大的银锁、戴银项圈，头顶银饰，全身的银饰加起来重8～10千克。

绉绣、平绣、堆绣是苗族特有的绣法，这些刺绣不同于其他的刺绣，所呈现的图案立体而惊艳。

With a population of more than 9 million, the Miao people form one of the largest ethnic minorities in southwest of China. Because of the great varieties and long history, the Miao costumes and ornaments have been honored as the “wearable history book”. Miao men usually wear linen jackets with colorful designs, and drape woolen blankets with geometric patterns over their shoulders, or wear short jackets buttoned down the front or on the left and long trousers and turbans. Miao women's dressing varies from village to village. In some areas, women wear jackets buttoned on the right and trousers with decorations embroidered on collars, sleeves and legs. In other areas, women wear full-or-half length or short pleated skirts. They also love to wear kinds of silver jewelry on festive occasions.

苗族银饰以大为美，以多为贵，历经千年，在历史人文、民族心理、生活美学方面都有较高的研究价值。
Miao people like silver ornament, and the bigger and the more, the better. Miao silver-work is important in research of historical humanity, ethnic mentality and life aesthetics.

苗族服饰是生动自然的生活装，一切纹饰都是发自内心的需要。
Miao costume is the real portrayal of nature and heart.

苗族日常便装也有和现代衣料不断融合之处，但整体风貌传承得很完整。
Contemporary Miao garments still maintain the modern style and traditional characters.

苗族小孩帽
Miao kids with traditional headgear

苗族锦鸡服极具现代艺术气质。

Miao Golden Pheasant Dress is full of modern artistic style.

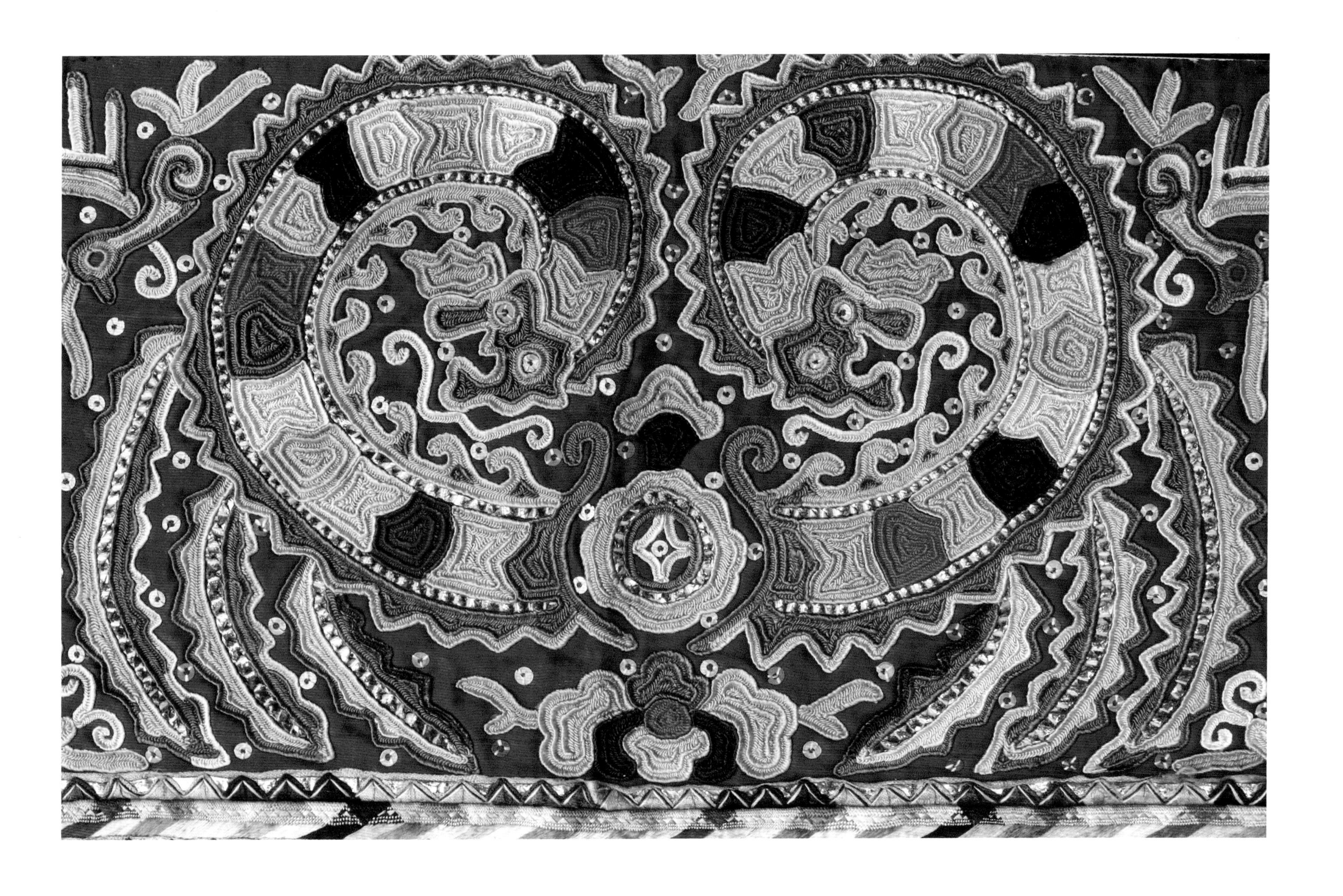

绉绣、平绣、堆绣是苗族服饰特有的绣法，所呈现的图案立体而惊艳。
With different stitch techniques and designs, Miao embroidery looks three-dimensional and colorful.

苗族
Miao

Miao costume is praised as the "wearable history book" that reflects this ethnic group's long history, their pursuit of life as well as their unique aesthetics. Miao people make wonderful embroidery, silver jewellery and batiks.

国家级蜡染工艺传承人王阿勇创作的花鸟蝴蝶纹蜡染作品

Batik work of butterfly and birds pattern by Ms. Wang Ayong, a national inheritor of traditional batik craft

彝族
Yi

彝族支系众多，居住分散，各地服饰区别明显，样式各异，带有浓厚的地域色彩。彝族服饰款式变化多姿，常以大量银制品和刺绣装饰。

彝族主要聚居在中国的四川省和云南省，人口共有871.4万人。彝族有本民族的语言和文字。

彝族传统服饰展现了畜牧文化特色，崇尚黑色，一直保持古朴、独特的民族风格。

男女都披“披毯”和“擦尔瓦”，昼为衣，雨为蓑，夜为被。男性着窄袖上衣，穿长裤，豪迈威武，表现其英雄气概；女性上衣配百褶裙，丰富多彩，彰显彝族人对美的追求。

The total population of the Yi ethnic minority is about 9 million. They live in Yunnan and Sichuan, Provinces in southwestern China. The Yi people's traditional costumes vary in different regions. Men usually wear black jackets with tight sleeves and right-side askew front, and pleated trousers. Women wear embroidered jackets and pleated long skirts hemmed with multiple layers. The Yi people prefer the black color. They wear turbans and take a cloak for all the seasons and weather.

彝族姑娘的高盘帽
A Yi girl wearing a traditional high hat

彝族妇女多着百褶长裙，用宽布与窄布镶嵌横联而成。
Pleated dress of Yi women

四川凉山彝族妇女的银质头饰
Silver head-wear of Yi women Liangshan area of Sichuan Province

云南省漾濞县富恒乡彝族服饰（男子穿羊皮褂，女人穿刺绣服饰）
Traditional Yi costumes, Yangbi County of Yunnan Province

彝族服饰
Traditional costumes of Yi ethnic group

精美的彝族帽子
Beautiful caps of Yi people

Because the people of Yi ethnic group scatter in different areas, their attires are divergent. The styles, designs and patterns vary from one locality to another. Yi people are fond of silver ornaments.

四川布拖彝族男女都喜欢用银制大扣来装饰自己。
In the Butuo tribe, Yi people like large-size silver buttons.

土家族

Tujia

土家族服饰以俭朴实用为原则，结构简单，但注重细节，喜宽松的衣服，衣短裤短，袖口和裤管肥大。织锦是土家族服饰的重要装饰。

土家族主要聚居在中国湖南、湖北、四川、贵州省，人口共有835.3万人。土家族有自己的语言，但无文字。

土家族传统服饰讲求简洁朴素，便于生产劳动。妇女擅长纺织刺绣，其中以西兰卡普织锦著称。相传，西兰卡普为一个名叫西兰的土家族姑娘所创，一般多用蓝色或黑色纱线为底纱，再用五颜六色的彩色丝织物组成各种精美的图案，图案有二三百种，用于服饰和被面。西兰卡普十分注重色彩的对比和反衬，强调一种鲜艳脱俗、清新明快、安定协调的视觉效果，体现土家族大方热情的民族特色。编织西兰卡普是土家族姑娘的必备本领，而西兰卡普又是土家族姑娘的嫁妆。

Most of 8.3 million Tujia people live in Hunan Province and the rest live in the neighbouring Sichuan, Guizhou and Hubei Provinces. Tujia people are very good at making their own brocade,called “xilankapu”. They use the brocade to decorate garments. Tujia men wear a coat and a collarless shirt. A long band is usually tied in the waist. Trousers, usually in green and blue, are fat with loose and short bottoms. Women’s costumes have many varieties in shirt, coat, gown, trousers and skirt. These clothes are beautifully embroidered and decorated with laces. They also like to wear earrings, and bracelets made of gold, silver and jade.

西兰卡普被称作“土家之花”，在土家族人民生活中有着实用的、礼俗的和审美的三方面意义，不仅以经久耐用著称，而且是土家族婚俗中的主要嫁妆。

Xilankapu brocade, the most famous handicraft of Tujia ethnic group, is not only famous for its outstanding durability, but is the principal dowry in Tujia wedding customs.

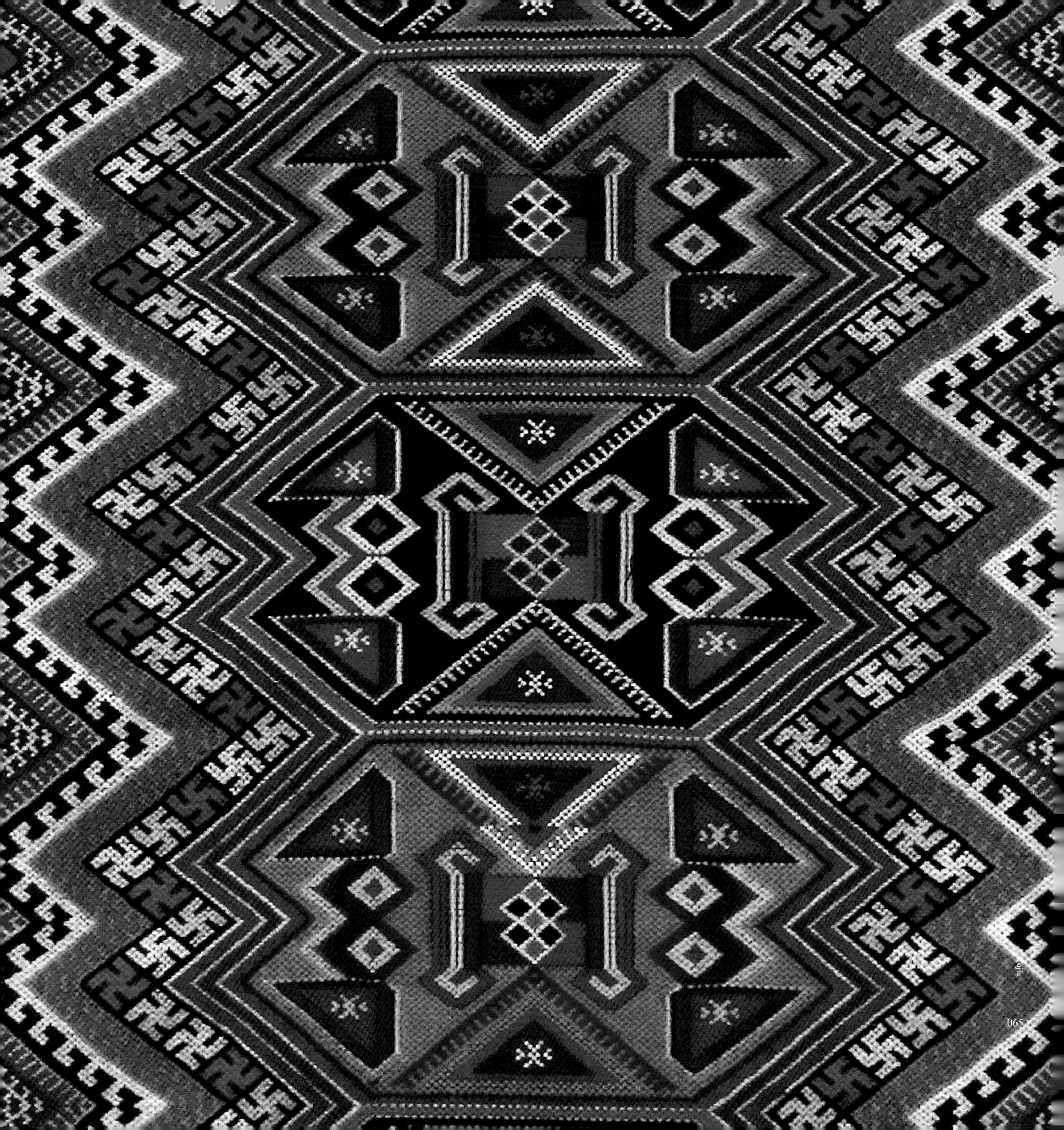

土家族
Tujia

People of Tujia ethnic group pay attention to detailed decorations on their loose garments. Tujia Xilankapu is the well-known brocade that is hand-woven by Tujia people.

土家族女服
Traditional upper-wear of Tujia women

藏族
Tibetan

藏族各地服饰有着不同的文化特色，但在色彩的认同上是一致的，这就是藏地五色——白、蓝、红、黄、绿。

藏族主要分布在中国的西藏、青海、甘肃、云南和四川等省区，人口共有628.2万人。藏族有本民族的语言（藏语）和文字（藏文）。

藏族传统服饰是肥腰、长袖、大襟长袍，夜晚睡觉当被子，冬季可以防寒保暖，夏季气温升高的时候可以脱掉其中一个袍袖，调节体温，方便散热。

藏族穿衣服喜欢搭配金、银、珍珠、玛瑙、松石、翡翠、珊瑚、蜜蜡、琥珀、丝绸、玉等材质的头饰、发饰、鬓饰、耳环、项链、胸饰、腰饰、戒指等。女性多佩戴珊瑚玛瑙饰坠、项链和银质佛盒；男性通常佩戴各式腰刀、火镰等饰物，同时也会佩戴耳环、戒指和手镯等。

藏族服饰面料材质众多，平日里穿着的藏袍采用面料为素皮面、白氆氇毛织物等；参加重大活动的时候，通常穿由提花皮面料、蟒缎袍、团花锦缎袍、云锦等织成的藏袍。

The Tibetans with a population of 6.28 million mostly live in the Tibet Autonomous Region. There are also Tibetan communities in Qinghai, Gansu, Sichuan and Yunnan Provinces. All Tibetans, men and women, like to wear ornaments. They wear fine fur hats, long-sleeved silk or cloth jackets topped with loose gowns which are very big and also serve as mattress and quilt at night. They often leave one or both arms uncovered while tying sleeves around the waist, making it convenient for work.

生活在西藏自治区阿里地区的妇女们在黄昏中起舞
Tibetan women’s attire

藏族妇女的帽饰
Tibetan women's headdress

云南香格里拉藏族男子服饰
Costumes of Tibetan men, Shangri-La, Yunnan Province

藏族妇女的发饰
Tibetan women's headdress

藏族妇女珍珠冠头饰
A Tibetan woman wearing a pearl and silver headwear

藏族

Tibetan

Although the costumes of Tibetan people vary from different areas, the colors they like are same: white, blue, red, yellow and green.

藏族男子的服饰

Tibetan men's costume

蒙古族

Mongolian

蒙古族服饰以袍服为主，便于鞍马骑乘，具有浓郁的草原生活特色。款式多褒衣博带，既展现了人体的曲线美，又能体现蒙古民宽厚大度、粗犷坦荡的性格。

蒙古族主要分布在中国的内蒙古自治区，是“逐水草而居”的游牧民族，人口共有598.1万人。蒙古族有自己的语言和文字。

蒙古族的传统服饰主要由长袍和隆重的头饰构成。蒙古袍四季都可穿，春天和秋天穿夹袍、夏天穿单袍、冬天穿棉袍或皮袍；袍装会佩戴珊瑚、玛瑙、翡翠、珍珠、琥珀、白银制成的头带、头圈、辫钳、辫套、项链、头簪、头钗、耳环、手镯、戒指等首饰。

鄂尔多斯地区的蒙古族头饰有“头饰之冠”的美称，由各种银饰、簪花吟片、红珊瑚等镶嵌而成，制作工艺流程复杂，高贵典雅。

The Mongolians, with a total population of over 5.98 million, live mostly in the Inner Mongolia Autonomous Region and the rest reside in many provinces in China. Fur coats lined with satin or cloth or nothing at all in winter and loose, long-sleeved robes in summer, the Mongolians like this kind of gowns all the year around. The Mongolian costume is generally red, yellow or dark blue in color. A red or green waistband, flint steel, snuffbox and knife are accessories to both men and women. Knee-high felt boots are a typical common footwear. They wear cone-shaped hats, silk or cloth turbans. They like to decorate their dresses with agate, coral and green jade.

参加那达慕的布里亚特蒙古族妇女
Mongolian women wearing costumes for Naadam Fair

鄂尔多斯女性头饰

Mongolian headwear, Erdos, Inner Mongolian Autonomous Region

蒙古族礼服
Mongolian costumes for ceremony

内蒙古自治区2018年搏克大赛

蒙古族
Mongolian

Mongolians like to wear gowns and robes because this style of dress is good for riding and demonstrates their character of boldness and vigor.

蒙古族摔跤服图案粗犷有力，色彩对比强烈，各部分配搭恰当，具有勇武的民族特色。
Mongolian wrestling costume

侗族

Dong

侗族大都穿自纺、自织、自染的侗布，喜青、紫、白、蓝色。其服饰千姿百态，平时穿着便装，讲求实用，盛装时注重装饰，朴素与华贵相得益彰。

侗族主要分布在中国的湖南、贵州、广西壮族自治区，人口共有287.9万人。侗族有自己的语言，新创了侗文。

侗族传统服饰基调为青、蓝、紫、白等色，点缀浅绿、浅玫瑰红，色彩冷暖互补，因地域分布呈南北两类：南部男性上衣有对襟、左衽、右衽三种，下着长裤，裹绑脚；女性下装分裙、裤两类，穿裙时，上身以开襟紧衣服相配，胸部围青色刺绣的剪刀口状“兜领”，裹绑脚；穿裤时，以右衽短衣相配。北部女性上衣右衽无领，以银珠为扣，环肩镶边，裤及膝下12厘米左右，足蹬翘尖绣花鞋，腰系飘带。

侗族的服饰特色面料是亮布，是一种经过染色的粗布料。它的制作要经过浸染、捶布、晾布和涂抹鸡蛋清等十多道工序，工艺十分古朴，穿着经久耐用，别具民族特色。这种土布晒干后闪闪发亮，所以俗称“亮布”。

The population of the Dong ethnic group is about 2.8 million, living in the extensive stretch of territory hills on the Hunan-Guizhou-Guangxi borders in southwest China. The garments they wear vary by different areas. Men usually wear short jackets with front buttons. In the mountainous localities of the south, they wear collarless skirts and turbans. Women are dressed in skirts or trousers with beautifully embroidered hems. Women wrap their legs and heads in scarves, and wear embroidered shoes.

侗族无领大襟衣
Dong Collarless jacket

侗族绣花鞋
Dong shoes with embroidery

贵州丽萍肇兴侗族大歌
Dong costumes in Zhaoxing, Guizhou Province

贵州肇兴男子服装的后背上绣满龙凤和花草
Dong men's costumes full of embroidered dragon, phoenix, flowers and plants in Zhaoxing, Guizhou

戴绣花头帕的侗家女
Dong girls with embroidered headscarf

侗族
Dong

Dong people like to wear costumes of self-made cloth in blue, purple, white and green. Their everyday wear being simple and practical, they will fully dress up for festivals.

身穿节日盛装参加“祭萨”的侗族妇女
Dong women in festival costume for “Jisa”(sacrifice)

布依族

Bouyei

布依族服饰大体上都保留着古老的特点。女衣裙均有蜡染、挑花、刺绣图案装饰，既庄重大方，又新颖别致。

布依族主要聚居在中国的贵州省，是中国古越人的后裔，人口共有 287 万余人。布依族有自己的语言，后创布依文。

布依族传统服饰制作集蜡染、扎染、织锦、刺绣等工艺技术于一身，其图案多为蕨菜花、刺梨花、团花、小碎花，仿效铜鼓纹的漩涡纹、水波纹；还有远洋、梅花、龙凤、蜡裂的冰纹，反映布依族独有的审美特征。

布依族采用古老的扎染方式，把织好的布折叠成各种图案，用麻线扎好进行浸染、漂洗，最后成为蓝底白花的各种图案。而他们的蜡染技术，远在宋朝就已经盛行。

Most of China's over 2 million Bouyei people live in areas in Guizhou Province. The Bouyeis are skilled in art and craft. Their colorful and beautifully-patterned batik dates back to ancient times.From their costumes one could see not only batik but also embroidery and satin in a great variety of patterns. The Bouyeis are special in making beautiful garments.

在河边清洗蜡染布的布依族女孩
Dong costumes in Zhaoxing, Guizhou

布依族百褶裙
Bouyei pleated skirt

布依族
Bouyei

The characteristic of Bouyei costume is very traditional. With batik and embroidery, the Bouyeis are special in making beautiful garments.

瑶族
Yao

瑶族支系众多，各支系服饰也不尽相同。瑶族曾因服饰的颜色、裤子的式样、头饰的装扮不同而有不同的族称。

瑶族主要聚居在中国的广西壮族自治区，人口共有279.6万人。瑶族支系繁多、分布广，按照语言、习俗和信仰等方面的差异一般划分瑶语支、苗语支、侗水语支、汉语支四大支系。

瑶族各支系的服饰存在差异，同一支系的瑶族因居住地不同，服饰也不尽相同，如广西南丹瑶族男子穿交领上衣，着白色大裆紧腿及膝短裤，故得名“白裤瑶”。

瑶族妇女在印染实践中摸索出一套完整的蓝靛印染技术。她们从自己种植的蓝草中提取蓝靛，加入白酒，用草木灰过滤，发酵至黄色便可染布。

The Yao people, with a population of about 2.8 million, live mainly in the Guangxi Zhuang Autonomous Region. Some of them live in the neighbouring provinces. Men like to wear front-opening short robes without collars, together with long pants or knee-length shorts. Women like to wear side-opening collarless jackets and long pants, shorts or accordion pleated skirts.

广西贺州过山瑶妇女的服饰色彩极其丰富
Yao women's costumes are very bright and colorful, Hezhou, Guangxi.

绣花帽
A Yao woman is embroidering a hat.

广西贺州黄洞瑶族
Yao people's headdress, Huangdong Township, Hezhou city, Guangxi Province.

跑纱的白裤瑶妇女
A Yao woman is spinning.

瑶族
Yao

There are many tribes in Yao ethnic group. The costumes vary in different areas. The most eminent differences lie in color and head wear.

广西金秀瑶族男子服饰
Yao men's costumes, Jinxiu, Guangxi

白族
Bai

白族崇尚白色，衣物以白色为贵，再配以色彩对比明快而映衬协调、挑绣精美的披挂，充分反映了白族人在服装艺术上的智慧。扎染是白族服饰的特色。

白族主要聚居在中国的云南省，人口共有193.3万人。白族有自己的语言（白语）和文字（新白文）。

白族尚白，传统服饰中男性身着白色上衣和黑领褂，下穿白色长裤，头缠白色或蓝色头帕，肩挂绣着美丽图案的挂包；女性多穿白色或浅蓝色上衣，外套红色或黑色丝绒领褂，右衽结纽处系银"三须""五须"饰物，下着白色或浅蓝色宽裤，腰系绣花或缀有绣花飘带的短围裙，足蹬绣花鞋。

在白族姑娘的头饰上，蕴含着一个大家非常熟悉的词语，它就是"风花雪月"：垂下的穗子代表着下关的风；艳丽的花饰代表着上关的花；帽顶的洁白代表着苍山的雪；弯弯的造型代表着洱海的月。

Bai people mainly live in Yunnan Province, with a population of 1.93 million. They have their own language and writing system.

Bai people adore white. In traditional clothing, men usually wear a white jacket with a black-collar garment, white trousers, white or blue headscarf, and a beautiful-patterned bag hanging across the shoulder; women wear white or light blue jacket, red or black velvet coat, white or light blue loose trousers, a short apron with embroidery or with embroidered ribbons, and embroidered shoes.

Dali girl's headwear is rich in connotations, wind, flower, snow and the moon. The drooping tassels represent the wind in Xiaguan(very strong); the gorgeous flower decoration, the flowers in Shangguan; the white of the headwear's top, the snow of Cangshan mountain; the arched shape, the moon above the Erhai Lake.

白族妇女的扎染技术相当高超
The technique of batik is superb among Bai women.

与大理白族妇女身上那代表着“风花雪月”的穿戴截然不同，甸南妇女的服装款式宽松，以黑、红为基本色调，乍一看与清朝的宫廷服饰有异曲同工之妙。

Different from the Bai people in Dali who like white, the Bai ethnic women in Diannan area wear costumes in black and red.

白族
Bai

The main color of costumes of Bai ethnic group is white. They are quite good at making costume decoration. Batik is one of the characters of Bai costumes.

朝鲜族
Korean

朝鲜族自古有“白衣民族”之称，认为白色纯洁自然；同时，朝鲜族妇女们的衣料颜色也是绚丽多彩、不拘一格，但短衣长裙这一传统民族风格久久不变。

朝鲜族一个世纪以前从朝鲜半岛移居到中国东北地区，人口共有183万余人。朝鲜族有本民族的朝鲜语和朝鲜文。

朝鲜族的传统服饰有男女上衣、坎肩、裤子、裙子、外套、笠帽。男性多穿宽腿、肥腰、大裤裆的长裤，裤腰多余的部分折叠在前面，再用腰带系上，裤筒下也用布带系上；上衣则为短款式样，上衣外面还喜欢套上背褂，颜色多为黑色。女性穿着为短衣长裙，这也是朝鲜族妇女服装的一大特色，象征温顺、善良和勤劳淳朴的美德。女性服装根据不同年龄选用不同的款式和布料，其中儿童时期的衣服颜色色彩斑斓，通常有七种颜色，俗称“七色缎”。

There are about 1.8 million Koreans living in the northeast part of China. The traditional Korean dress is white, a symbol of simplicity and serenity. Men wear baggy trousers fastened at the ankles and a jacket fastened on the right; sometimes they wear a high-crowned black horsehair hat. Women wear voluminous robes and tight jackets which reach just below the armpits. The garments for children are colorfully designed.

吉林延边朝鲜族儿童服饰
Korean children in traditional costumes, Yanbian, Jilin Province

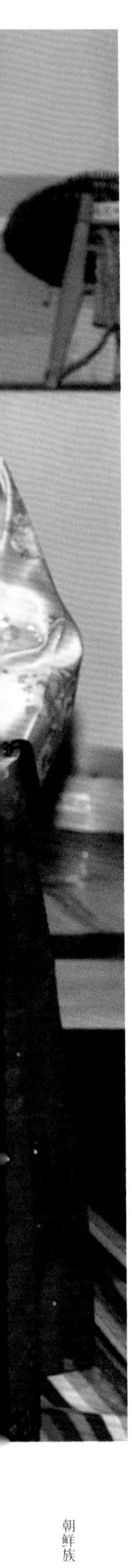

朝鲜族
Korean

Korean people like wearing clothes in white as well as colorful costumes. A short upper wear together with a long robe is the traditional style that remains unchanged for many years.

哈尼族
Hani

哈尼族服饰不仅仅是御寒防风蔽身之物，它还承载着极其丰富的文化信息。其服饰色彩、款式和纹样，既是哈尼族生存环境的反映，也是本民族社会身份和角色的标识。

哈尼族主要聚居在中国的云南省，人口共有166万余人。哈尼族有本民族的语言，1957年创建了一种以拉丁字母为基础的哈尼文字。

哈尼族崇尚黑色，擅长用蓝靛染布，他们的服饰布料多采用自己染织的靛青色土布。饰物多是银制品，如纽扣、耳环、项圈和手镯等。佩饰是哈尼族服饰的重要组成部分，主要佩饰有挎包、项圈、手镯、耳环、银链、银梳、银币、银铃、银泡、银针筒、帽子、围腰、腰带等。佩饰不仅多种多样，质地也各异。

银泡缀饰是哈尼族服饰一个引人注目的亮点，除普遍使用的银饰品外，海贝、羽毛、料珠、毛线、缨穗、骨针、绿壳虫等也用来作装饰。在黑色的底布上镶钉银泡花纹图案，显得华丽而朴实。

The Hani ethnic minority, with a population of over 1.6 million, live in the river valleys in the southern part of Yunnan Province. The Hani people prefer clothing made of home-spun dark blue cloth. Men wear front-buttoned jackets and trousers, and black or white turbans. Women have collarless, front-buttoned blouses with cuffs and trouser legs laced. They also wear skirts, round caps, and string of silver ornaments. Both men and women wear leggings.

The costumes of Hani ethnic group are both the clothing to wear and also passing cultural signals to show the difference in area and social status.

云南西双版纳哈尼族的头饰用小银泡和银币状的银片做装饰
Hani people's silver headdress

哈尼族木屐舞
Hani clog dance

黎族
Li

黎族因方言地域、族源、族系以及生活环境的差异，服饰的款样标准也不相同。黎族所织的"黎锦""黎单""黎幕"，色彩鲜明，美观实用。

黎族生活在中国的海南省，是中国南方一个古老的民族，人口共有146.3万人。黎族有自己民族的语言，1957年创制了以拉丁字母为基础的黎文。

黎族妇女擅长纺织，以黎锦著称。传统服饰中最具特色的是女性短裙。因终年跋山涉水，上坡下田劳作，天气湿热，黎族女装逐渐形成了五寸超短裙特色。

此外，黎族服饰以蓝黑色为主色，男性通常穿着无领无扣的长袖上衣，下穿左右开衩的布裙；女性上衣配花筒裙，头发挽髻包头或缠头。

With a population of 1.4 million, the Li ethnic minority live in Hainan Province, China's second largest island after Taiwan. The Li people prefer clothing in blue and black. Men wear collarless jackets and two pieces of cloth instead of trousers. Women wear buttonless blouses and tight-fitting long skirts.They do their hair in a coil behind and pin it with bone hairpins and wear embroidered kerchiefs.

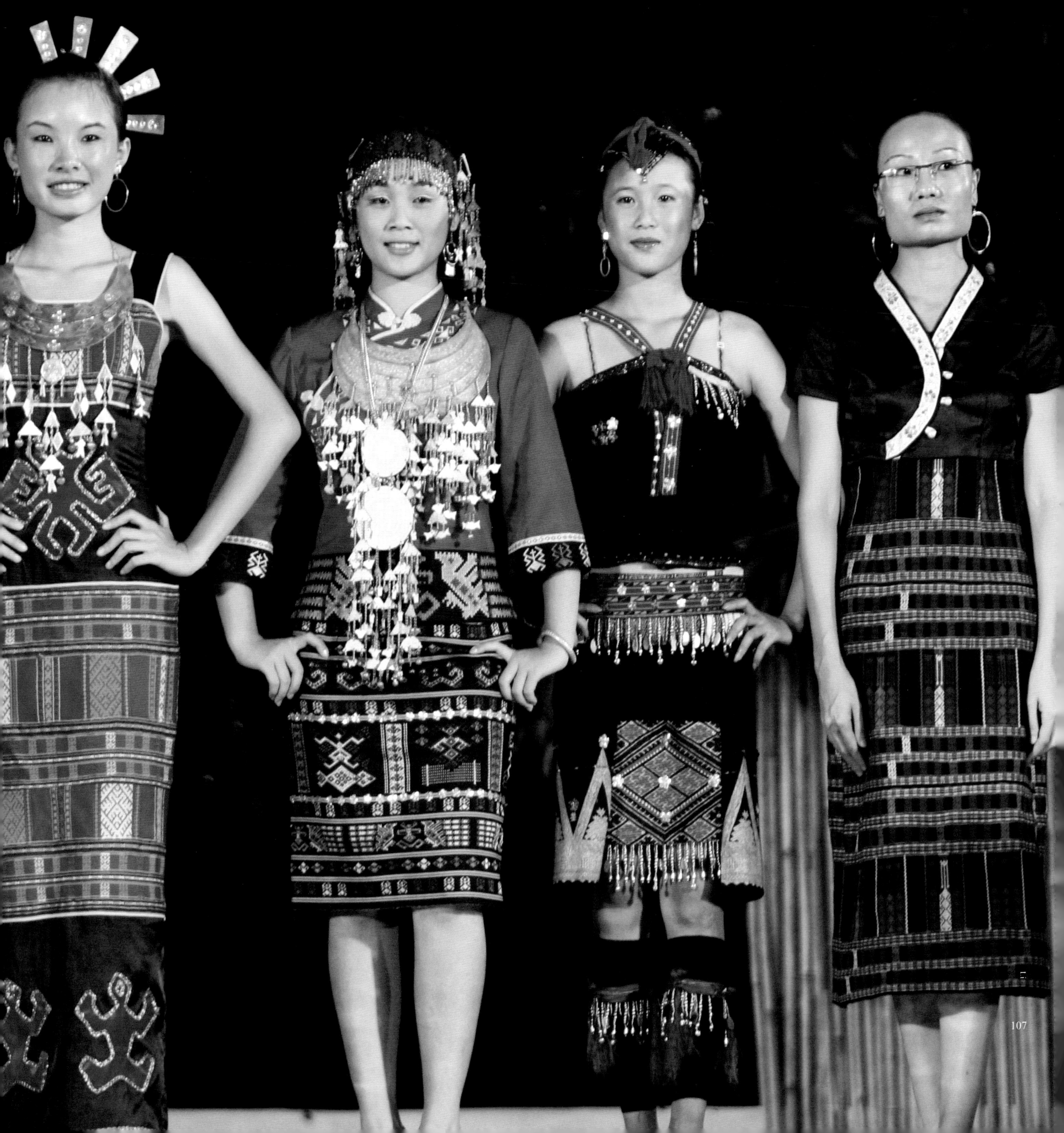

黎锦至今已有近 3000 年历史，所传承的花纹图案堪称黎族的独特史书，是中国乃至世界上最为古老的棉纺织染绣技艺。

The Li brocade has a history of almost 3,000 years. The patterns and designs, plus weaving techniques, reflect one of the most ancient cotton weaving and dyeing skills.

哈萨克族
Kazakh

哈萨克族是以草原游牧文化为特征的民族，服装便于骑乘，服装多用羊皮、狐狸皮、鹿皮、狼皮等制作，反映了山地草原民族的生活特点。

哈萨克族主要聚居在中国的新疆维吾尔自治区，人口共有146.2万人。哈萨克族有自己的民族语言，使用阿拉伯字母文字。

哈萨克族妇女爱穿连衣裙，但最讲究的是头饰，未出嫁的姑娘一般戴头巾或是戴一种下沿大、上沿小，呈圆斗形的帽子，多用红色和绿色的绒布制作，用金丝绒线绣花，并用珠子镶成各种美丽的图案，帽顶上插一撮猫头鹰的羽毛，象征勇敢、坚定。女性出嫁需头戴由毛毡制成，外饰布缎金银珠宝，绣花的“沙吾克烈”尖顶帽。

The Kazakh ethnic minority, with a population of about 1.5 million, live in the Xinjiang Uygur Autonomous Region. The horse-riding Kazakh herdsmen are traditionally clad in loose, long sleeved furs and garments made of animal skins. Women like to wear embroidered cloth with silver ornaments. Both men and women like to wear animal-skin caps.

哈萨克族姑娘出嫁时爱穿白色的礼服
Bride of Kazakh

哈萨克族男子服饰
A Kazakh man in traditional costumes

62
6

哈萨克族妇女刺绣
Kazakh women doing embroidery

哈萨克族
Kazakh

The horse-riding Kazakh herdsmen are traditionally clad in loose and long-sleeved furs and garments made of animal skins.

哈萨克族男子服饰
Costumes of Kazakh men

傣族
Dai

傣族服饰淡雅美观，既讲究实用，又有很强的装饰意味。傣族织锦，历史悠久，图案丰富多彩，具有浓厚的生活色彩，色调鲜艳，风格纯朴。

傣族主要聚居在中国云南省，人口共有126.1万人。傣族有自己的语言和文字。

傣族传统服饰，男性多穿无领短袖，肥筒长裤，用白、青、浅蓝、淡黄色的布包头；女性上穿白色、绯色或淡绿色紧身窄袖（或长或短）短衫，下着各式花纹的长及脚面的筒裙，束银腰带。

傣族崇拜孔雀、大象，常将孔雀、大象的图案编织在筒裙等衣物上，如此的服饰就像一部记录地域风情的史书。

The total population of the Dai ethnic group is 1.26 million and they mainly live in Yunnan Province.

Traditionally men wear tight-sleeved collarless jacket and long, loose trousers. Headgear includes white, black and blue turbans. The Dai women prefer a short upper-wear and a long narrow skirt with a silver bend around waist.

Because the Dai people worship elephant and peacock, the similar patterns are used frequently on their costumes.

傣族妇女服饰

A Dai girl in traditional costumes

傣族
Dai

The Dai costumes are simple and elegant in color, practical to wear. The Dai brocade is colorful and rich in patterns. It is designed to be close to daily life and the style is simple yet elegant.

傣族泼水节
People wearing traditional costumes are celebrating the Water Splashing Festival.

畲族

She

福建省畲家妇女服饰以象征万事如意的“凤凰装”最具特色，尤以发式、发饰别具一格。

畲族主要聚居在中国的福建省和浙江省，人口共有70.8万人。畲族有自己的语言，但没有文字，通用汉文。

畲族传统服饰崇尚蓝色。男性穿青色圆领短衫阔袖麻布衣，年老男性头扎黑布头巾。女性传统服饰最具特色的是“凤凰装”；衣服和围裙上刺绣彩色花纹，镶嵌金丝银线，象征凤凰的颈、腰和羽毛；盘起高髻扎红头绳垂缨，象征凤髻；全身佩戴银质品叮当作响，象征凤凰鸣叫，生动形象。

The total population of the She ethnic minority is about 700,000. They are scattered in China's southeast provinces, mainly in Fujian and Zhejiang Provinces. The traditional costumes and accessories of the She ethnic minority are multicolored. She people like wearing green and blue, and materials are usually home-woven flax. Middle-aged and old men wrap their heads with black scarfs. The She women's phoenix coronet is the most distinctive dress. With a round silver plate and three smaller silver plates, a silver hairpin is inserted on the coronet. The accessories such as silver necklace, silver bracelet, silver chains and silver earrings are worn, which are bright-colored and shiny.

浙江景宁畲族婚礼
A Wedding of the She people, Jingning, Zhejiang Province

织带是畲族古老的手工工艺

Belt weaving is She people's handcraft inherited from their ancestors.

畲族
She

Women's costumes of the She ethnic minority are very uniquely characterized, especially the hair style and headdress.

福建畲族妇女便装
She women's casual wear in Fujian Province

畲族女子头饰
She women's headdress

傈僳族
Lisu

傈僳族因服饰颜色的差异分为白傈僳、黑傈僳、花傈僳。

傈僳族是跨界民族，国内主要分布在云南省和四川省，人口共有70.2万人。傈僳族有本民族的语言和文字。

傈僳族生活在山林中，于是就地取材，用山中的棕榈、火麻等为原材料织布纺衣。

傈僳族传统服饰中，男性着大襟麻布过膝长衫，腰系布带，装饰贝壳；女性衣裙用白片各色布料精制而成百布衣，外罩黑绒褂子，下穿多褶麻布长裙，头戴“俄勒”，胸前佩戴银饰。

The Lisu people, with a population of 702,000, live in areas in the northwestern part of Yunnan Province. The Lisu people wear clothes woven by themselves. In the traditional style, men wear linen gown or jacket and long-to-knee pants. Some of them wrap their head with black cloths and some with their long hair plaited on the back of head. Lisu woman's dress is of great varieties on the occasion of festival and wedding, it is dazzling to see the girls wearing colorful clothes, such as the long robe with colorful laces in the collar, breast and cuff. It is made of twelve 5-inch-wide cloths of 5 colors. The Lisu women like to wear silver ornaments on the breast.

不同地区的傈僳族妇女因服饰颜色的差异而被称为白傈僳、黑傈僳、花傈僳。白傈僳妇女普遍穿右衽上衣、素白麻布长裙，戴白色料珠；黑傈僳妇女多是右衽上衣配长裤，腰系小围腰，缠黑布包头，戴小珊瑚之类的耳饰；花傈僳妇女喜穿镶彩边的对襟坎肩，搭配缀有彩色贝壳的及地长裙，缠花布头巾，耳坠大铜环或银环，摇曳多姿，风情万种。

Since the Lisu women of different areas prefer different colors, they are identified accordingly. They also dress differently. There are three main groups, black Lisu, white Lisu and flowery Lisu.

劳作中的傈僳族妇女
Lisu women at work

傈僳族
Lisu

The Lisu ethnic groups are divided according to the colors of their costumes and accessories.

东乡族
Dongxiang

东乡族至今仍保留着不少古代游牧民族的传统和生活习惯。

东乡族主要聚居在中国的甘肃省，人口共有62.1万人，有本民族的语言，无本民族文字。

东乡族聚居的地区，处于青藏高原与黄土高原的过渡地带，这为东乡族传统服饰提供了天然的原料——羊皮毛。羊毛制品是东乡人生活服饰中的重要内容。

东乡族的服饰特点也颇为明显。女性喜穿圆领大襟宽袖绣花衣服，颜色以青黑为主。由于东乡族女性喜爱刺绣，下装和鞋上常有刺绣装饰。不同年龄分别穿戴绿、黑、白三色盖头。男性上穿大长袍束腰带，挂腰刀荷包，戴“礼拜帽”。

The population of Dongxiang ethnic minority is about 621,000 and they live mainly in Gansu Province and Xinjiang Uygur Autonomous Region. Traditionally, sheep skin was the main material for their clothing. Women like embroidered jacket and trousers as well as shoes. Men wear gowns with belt and sword. Women wear veils in green, black or white according to ages while men wear white hats without brim.

The ancient nomadic living habit and tradition can be seen in the daily life of the Dongxiang ethnic group.

东乡族男子的礼服“仲白”是上清真寺礼拜五聚礼的礼服，须经常保持洁净。
Dongxiang men's costumes for religious ceremony

仡佬族
Gelo

仡佬族人善纺织、刺绣、蜡染，历史上因其服饰色彩款式不同而被称为“青仡佬”“红仡佬”“花仡佬”“披袍仡佬”等。仡佬族以精工巧织的“铁笛布”而闻名。

仡佬族主要聚居在中国的贵州省，是中国西南地区一个古老的民族，人口共有55万余人。仡佬族有自己的民族语言，无文字。

仡佬族的传统服饰，男性多穿着青蓝色无领上衫，系布纽扣，长衣袖系长腰带，下着接腰大脚裤，绑腿，头包青色头帕；女性上着镶彩色花边上衣，下穿镶花边裤子。银饰是仡佬族女性的最爱。

More than 550,000 Gelos live in dispersed communities in western Guizhou Province and the neighbouring provinces.

Traditionally, men wear collarless jackets reaching to the knee with cloth buttons in the front in black or blue. They wear baggy trousers and leg wrappings. Women like to wear jackets and trousers with laced edges. They are fond of silver ornaments.

仡佬族全家福
A Gelo family photo

仡佬族
Gelo

The people of Gelo ethnic minority are good at textile, weaving, embroidery and batiks.

拉祜族

Lahu

拉祜族服饰大多以黑布衬底，用彩线和色布缀上各种花边图案，再嵌上洁白的银泡，服饰色彩对比浓烈。

拉祜族主要聚居在中国的云南省，人口共有48.5万人。拉祜族有本民族语言和文字。

拉祜族传统服饰反映了游牧文化和农耕风格，古代男女均穿袍服，挎自织镶嵌贝壳、绒球的长方形背带，喜欢戴银质项圈、耳环、手镯，女性胸前挂大银泡。崇尚黑色，以黑为美，头包4米长的黑色包头，在包头两端缀以线穗，有的则是包大毛巾。男性多穿圆领对襟短衣服，下穿斜拼裆宽管长裤；女性多穿右襟黑布两边开衩的长衫，下穿筒裙。

The Lahu ethnic minority has a population of 485,000,mainly distributed in Yunnan Province. Lahu men wear a collarless jacket buttoned on the right side, long baggy trousers, and a black turban. The women wear a long robe with slits along the legs. Around the collar and slits are sewn broad strips of color cloth with beautiful patterns and studded with silver ornaments.

拉祜族

Lahu

The costumes of Lahu people are characterized by sharp contrast of color in black cloth with white and colorful patterns.

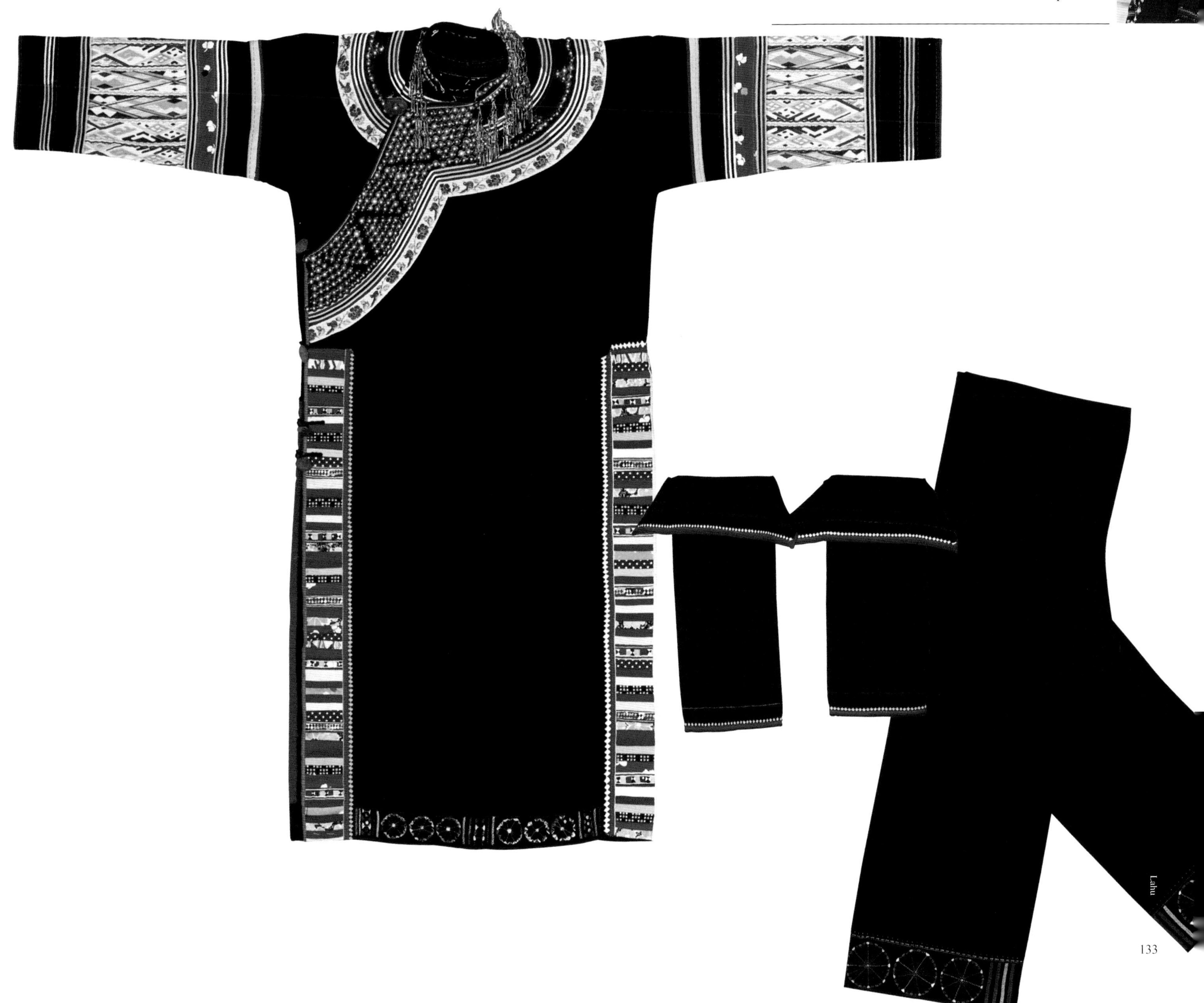

佤族

Va

佤族崇尚红色和黑色，服饰多数以黑为质，以红为饰，基本上还保留着古老的山地民族特色，显示着佤族人粗犷、豪放的坚强性格。

佤族生活在中国的云南省，是中国西南地区古老的民族，人口共有 42.9 万人。佤族有自己的民族语言和文字。

佤族服饰通常用自纺自织的棉麻为衣，黑色为主辅以红、蓝、紫、白、黄等色。最有特色的女性穿着是无领小坎肩配黑、红、黄色条纹筒裙，小腿裹护腿布，戴大耳筒，宽手镯、体现了佤族豪放坚强的性格，而男性多穿无领长袖衣，下穿肥大短裤。

With a population of over 400,000, the Va ethnic minority live in areas in southwestern Yunnan Province. The Va people make their clothing with cottons and linens which are grown by themselves. Men's garments usually consist of collarless jacket and wide trousers. Women wear a black short dress and a straight long skirt with folds. Va women are fond of bracelets round their wrists and earrings.

云南西盟佤族妇女服饰
The Va woman's costumes

云南西盟佤族男子服饰
The Va man's attires, Yunnan Province

云南西盟佤族女子服饰
The Va women in traditional costumes and headgear

节日期间，正是佤族男女青年们谈情说爱的好时节，喝下几杯水酒，才会说出心里话。
Festival is a good chance for young people to court his or her lover.

佤族

Va

The Va people like red and black colors for clothing.They use red materials as decoration on the black clothes. The design style demonstrates their tradition and characters.

佤族妇女头饰

A Va woman's headdress

水族
Shui

水族崇尚黑色和藏青色，忌大红、大黄的热调色彩，而喜欢蓝、白、青三种冷调色彩，反映了独特的服饰审美观。

水族主要聚居在中国的贵州省，人口共有41.1万人。水族有本民族的语言和传统文字。

水族人擅长刺绣，所绣马尾绣风格独特。在白色马尾上缠绕白丝线，以其为工具勾勒图案，再用彩色线以刺绣的方式填充走在中间，使得整幅图案色彩绚丽、形象逼真、结构完整。用这种方法绣成的“马尾绣背带”“绣花鞋”等在国内外都享有盛誉，也是他们服饰的重要组成部分。

With a population of 411,000, the Shui ethnic minority live in Guizhou Province. In traditional style, the elderly men wear a white short jacket topped with a blue robe and loose trousers. Women like to wear a long coat and trousers, aprons and embroidered shoes. They wear silver ornaments on festival occasions.

贵州水族妇女喜着绿色上衣
Green is Shui women's favorite color for clothing.

在水族地区，婴儿是在这样漂亮的马尾绣品襁褓中长大的，背带是外婆送给外孙的重要礼物，其上绣有各种寓意孩子健康成长的图案，一般一生只送一次。一副结实耐用的背带可供两三代人使用。
The embroidered swaddle strap is the precious gift to a newly born baby by the grandma on the mother's side. All the auspicious patterns are exquisitely embroidered. In the Shui area, this swaddle strap could be used for two or three generations.

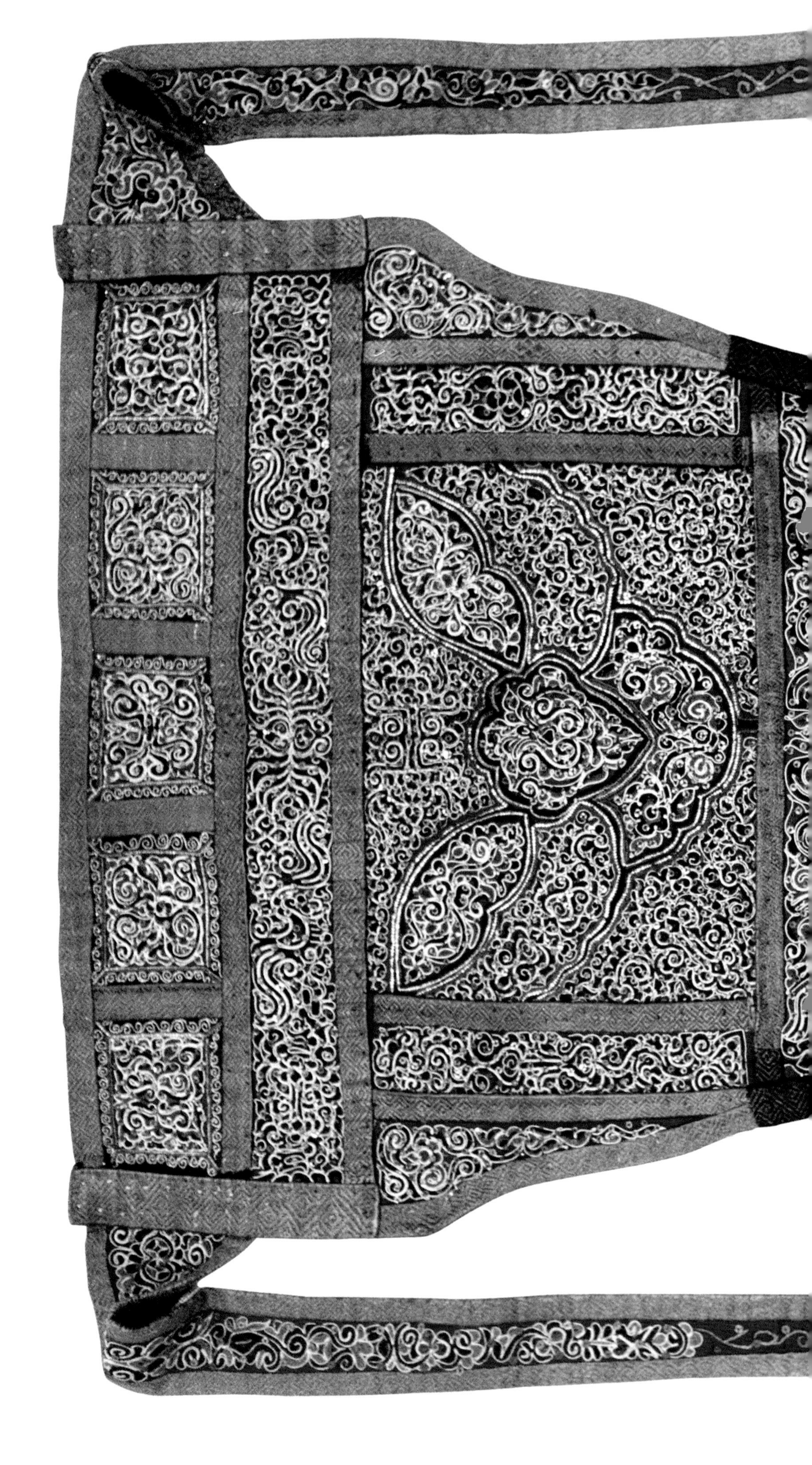

水族蝶纹马尾绣背扇
Embroidered back cover

水族
Shui

The Shui people are fond of black and dark blue colors. They use cool colors, blue, white and green-blue, instead of warm colors, red and yellow.

卯坡上对歌的水族妇女
Shui women in traditional-style clothes

纳西族
Naxi

纳西族妇女披肩上的七个刺绣圆盘象征肩担日月，背负星星，俗称“披星戴月”，象征着纳西族的勤劳。

纳西族主要聚居在中国的云南省，人口共有32.6万人。纳西族有自己的民族语言和文字。

纳西族传统服饰中，最具特点的是纳西妇女的七星羊皮披肩，上面缀有两个大的圆形绣花、七个较小的圆形图案。关于七星的由来：传说古时候有个叫英古的姑娘，一心为百姓除去天上炙热的九个太阳。龙三王十分感动，便将九个太阳吞掉了八个，又吐出一个，变成了一个太阳和一个月亮，其余的七个太阳变成星星镶在了英古姑娘的披肩上。

The Naxi ethnic minority has a population of 326,000, most of whom live in a Naxi Autonomous County in Yunnan Province. Naxi women wear wide-sleeved loose gowns, with jackets and long trousers, tied with richly decorated belts at the waist. They often wear sheepskin slung over the shoulder, on which the embroidered patterns are really interesting. There are two large round patterns and seven smaller patterns. These patterns are considered as the symbol for hard-working and brave life of the Naxi people.

纳西族
Naxi

On the cape, Naxi women embroider two round patterns, which represent the sun, the moon and seven smaller patterns, which represent the stars. That is considered as a symbol of diligence of Naxi people.

羌族
Qiang

羌族是中国最古老的民族之一，其服饰、刺绣文化内涵深沉厚重、古朴神秘，对其周边的一些少数民族有着深远的影响。

羌族生活在中国的四川省，人口共有30.9万人。羌族有本民族语言和文字。

羌族无论男女都喜爱青色和白色头帕，穿麻布长衫，束腰带裹绑腿，套双面穿的羊皮褂子，晴天毛朝内，雨天则毛向外，御寒遮雨。羌族妇女的衣服绣有花边，有的衣领上还镶一排梅花形图案的银饰，系绣花腰带。

早在明清时代，刺绣已在羌族地区盛行，后来挑花技艺也为羌族妇女所喜爱。其针法除挑花外，还有纳花、纤花、链子和平绣等。除了刺绣，绣有云卷的"云云鞋"也是羌族独特的服饰，鞋形貌似小船，鞋尖微翘，鞋底较厚，鞋帮上绣有彩色云纹和杜鹃花纹图案。

The Qiang ethnic minority has a population of more than 300,000 who mostly dwell in hilly areas, crisscrossed by rivers and streams, in Maowen Qiang Autonomous Prefecture of Sichuan Province. The Qiang people dress themselves simply but beautifully. Men and women like gowns made of gunny cloth, cotton and silk with sleeveless wool jackets. They like to bind their hair and legs. Women's clothing is laced and collars are decorated with plum-shaped silver ornaments.They wear sharp-pointed and embroidered shoes, embroidered girdles and earrings, neck rings and silver badges.

羌族妇女正在用手工编织腰带
Qiang women are making waistband by hand.

羌族
Qiang

The Qiang ethnic minority is one of China's most ancient minority groups and so is that of its costumes and attires. The Qiang costumes influence other minority groups in the area.

穿在身上的传统羌族刺绣图纹
Traditional Qiang costumes and embroidery

土族

Tu

传统土族妇女的头饰异常华丽，称为“扭达”，且有八九种之多，包括吐谷浑扭达、适格扭达、加斯扭达、雪古郎扭达、加木扭达、索布斗扭达等。

土族主要聚居在中国的青海省，人口共有28.9万人。土族有本民族的语言，1979年创制了土语文字。

土族服饰最有特色的是女性头饰。

刺绣是土族妇女的传统手艺。土族人的衣边、衣领、披肩、腰带、花兜、男子的烟袋套、枕头等衣物上，往往都以精美的刺绣、盘线制成。图案多为“八瓣莲花”“孔雀牡丹”，表现了土族人的聪明才智，也是土族人热爱生活的写照。

The total population of the Tu ethnic minority is 289,000. They live in the northwestern part of China, in Qinghai Province.

Men like to dress in cloth robes, putting on high-collared fur gowns with a waist belt. They often dress up felt hats with brocade brims. The Tu women's dress are strikingly unique and beautiful. The shirts they wear are delicately embroidered and colorful and these shirts are called rainbow dress.

土族的刺绣独具一格，不论绣什么图案，都用“盘线”绣成。“盘线”是土族特有的刺绣针法，同时运用两根针线，做工精致、复杂、匀称，绣出的图案美观大方，朴素耐久。
Embroidery made with the typical and unique stitch of the Tu ethnic group.

土族妇女刺绣

Tu women are embroidering.

土族

Tu

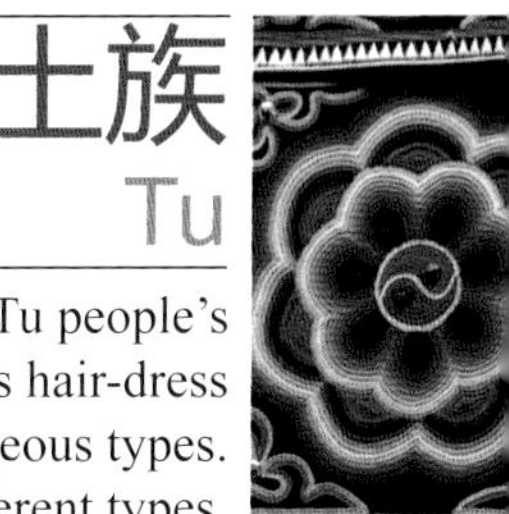

The most eminent feature of the Tu people's costumes and ornaments is the woman's hair-dress which is meticulously made into gorgeous types. There are totally nine different types.

土族刺绣带
Embroidery of Tu

土族千层底靴子
Handmade shoes of Tu people

仫佬族
Mulam

仫佬族服饰风格素朴简约。蓝靛染制的土布，仫佬人视为珍品，老年人的“防老衣”和姑娘们的“送嫁衣”都是用这种布料做成的。

仫佬族生活在中国的广西壮族自治区，人口共有21.6万人。仫佬族有自己的民族语言，没有文字。

仫佬族喜欢种棉花和蓝靛，织土布，自染，因此服饰喜欢青色。

仫佬族普通人家，成年男子一生之中只缝制一件长衫，用于做客时穿着；其他时间都穿无领短衫，长度可掩盖臀部，身宽袖大，前襟缝扣，开口于胸右侧，俗称“木桶盖”，也称“琵琶襟”。女性多穿无领及膝长衫，领口袖口缝三道“栏干”，梳辫挽髻，喜欢佩戴银饰。

The Mulam ethnic minority has a population of 216,000 and lives in the Guangxi Zhuang Autonomous Region. The Mulam People used to be famous for their spinning, weaving and dyeing. Deep blue is their favorite color. Traditionally, men wear jackets with large buttons in the front, long, baggy trousers and straw sandals. Women wear collarless long jackets to the knee with buttons in the front. Young girls braid their hair, which is coiled up onto their head after marriage. Women's jewelry includes silver earrings, bracelets and rings.

仫佬族
Mulam

The Mulam people are well known for cotton spinning, weaving and dyeing. They prefer to make garments with the cloth woven by themselves.

仫佬族男子服饰
Costumes of the Mulam men

锡伯族
Xibe

锡伯族早期服饰衣料以鹿、犴等兽皮为主。传统锡伯族服饰曾吸收过蒙古族、满族、汉族等民族服饰的优点。“套裤”（只有两条裤腿，没有裤裆和后腰）是锡伯族服饰的特色。

锡伯族主要聚居在中国的辽宁省和新疆维吾尔自治区，人口共有 19 万余人。他们使用自己的民族语言和本民族文字。

千百年来，锡伯族的祖先英勇善战，为捍卫边陲做出了贡献，而骑马射箭也延续至今，成为日常生活的一部分。为了方便骑射，他们的服饰十分考究，主要为左右开衩的长袍和马褂。

The Xibe ethnic minority, with a population of 190,000, are widely distributed over northern China from the Xinjiang Uygur Autonomous Region in the west to the northeast part of China. The Xibe women like the close-fitting long gown reaching the instep and big slits on both sides, and men wear short jackets with buttons down the front and trousers tightly tied around the ankle for convenient movement in fishing and hunting. They wear long robes in winter. They are fond of blue and black in color and women like colorful short jackets and scarfs.

锡伯族
Xibe

The Xibe People use animal skins for clothing. They also blend the elements of other ethnic groups' costumes into their own.

柯尔克孜族
Kirgiz

柯尔克孜族服饰的特点具有草原牧民的共性和本民族服饰特色。柯尔克孜人喜爱红色，其次是白色和蓝色。

柯尔克孜族主要聚居在中国的新疆维吾尔自治区，人口共有 18.6 万人。有自己的民族语言，使用阿拉伯字母文字。

柯尔克孜族的穿着中，最典型而又最普遍的是一年四季常戴、用羊毛毡制作的白毡帽。柯尔克孜族崇尚白色，以白色为纯洁的象征。关于白毡帽有个美丽的传说：过去他们的一位首领曾在远征之前召集四十位谋臣，下令统一军队的帽子。第四十位谋臣的女儿设计出了这款能躲避风雪又防止风沙的帽子，一直流传至今。

刺绣是他们服饰的特色工艺。

The Kirgiz ethnic minority, with a population of about 186,000, inhabit in the Xinjiang Uygur Autonomous Region and in Heilongjiang Province. Kirgiz men wear round-collared shirts trimmed with lace covered by a sheepskin jacket or a blue collarless, long gown. They wear loose trousers and high boots. Women wear loose collarless jackets with silver buttons down the front. The long and pleated skirt is bordered with fur. Some wear dresses pleated in the lower part, and covered with a black vest. They wear hats in green, purple or black. Unmarried girls wear their hair in many plaits, reduced to two after marriage.

柯尔克孜族妇女吹奏乐器"奥孜库姆孜"

A Kirgiz woman is playing a traditional music instrument.

柯尔克孜族男子一年四季多戴用羊毛制作的白毡帽，这是从衣着上区别柯尔克孜族的标志。
A Kirgiz man wears a white wool hat all the year round and that is the most remarkable sign for them.

柯尔克孜族
Kirgiz

The costumes of Kirgiz reflect the common elements of nomadic people and their specific characters. Their favorite colors are red, white and blue.

柯尔克孜族妇女头饰
The Kirgiz woman's hairdress

景颇族

Jingpo

景颇民族服饰主要有筒裙、头巾、头帕等。服饰风格粗犷豪放。

景颇族主要聚居在中国的云南省，人口共有14.7万人。景颇族有本民族的语言和文字。

在景颇族传统服饰中，男性穿黑色或白色短衣、黑色短裤，腰挎长刀，斜背挎包，头裹黑或白布；女性穿黑色短衣，下穿羊毛编织的鲜艳花筒裙，裹毛织护腿。最具特色的是衣领周围和胸前镶嵌银泡银链为装饰，通常挂六七串，走路的时候如铃铛一样叮叮作响。

The Jingpo ethnic minority, numbering 147,000, live mostly in Yunnan Province. Jingpo men usually wear black jackets with buttons down the front and short and loose trousers. Elderly people have a pigtail on the top of their head and covered with a black turban. Young people prefer white turban. They like to carry an embroidered bag, with a knife on the waist. Women wear short jackets and colorful tight skirts.

景颇族

Jingpo

The costume character of the Jingpo ethnic minority is vigorous and unconstrained. They wear black jackets and carry knives, women wear long narrow skirts, with scarf on the head.

盛装的景颇女子甩银袍

A Jingpo girl in her splendid attire is swithing her silver gown.

景颇族男子喜戴白、黑、蓝色的包头巾，包头巾的两端均缀饰有红、黄等色的小绒球。

The Jingpo men like to wear turbans in white, black or blue decorated with small red and yellow pompons on both sides.

达斡尔族

Daur

由于寒冷，达斡尔族以穿皮服装为主，服装以袍式为主。

达斡尔族生活在中国的内蒙古自治区和黑龙江省，人口共有13.1万人。达斡尔族有自己的民族语言，无文字。

由于达斡尔族一直生活在北方寒冷地区，因此他们的传统服饰以皮、布为衣，选皮与四季相宜，形成四季穿皮衣的服饰特点：秋末冬初制成的皮袍毛密而厚，抗寒保暖，结实耐磨；春夏和初秋猎获并制成的皮袍毛稀而短，皮质结实，具有抗湿防潮的功能；夏季则穿光板皮衣，清凉干爽。

There are over 131,000 Daur people living in the Inner Mongolia Autonomous Region and Heilongjiang Province, as well as in the Xinjiang Uygur Autonomous Region. The Daur People use deer skin hunted in the fall and winter for making their wears to keep warm while using those hunted in spring and summer for clothing in hot seasons. The women have always been known for their needle work, decorating their clothing with fine patterns.

达斡尔族妇女擅长手工刺绣，服饰、鞋、荷包等多绣着各种花纹及图案。

The Daur woman is good at embroidery.

达斡尔族
Daur

Because of the environment, the Daur people use deer skin for clothing and they wear gowns made out of animal skins.

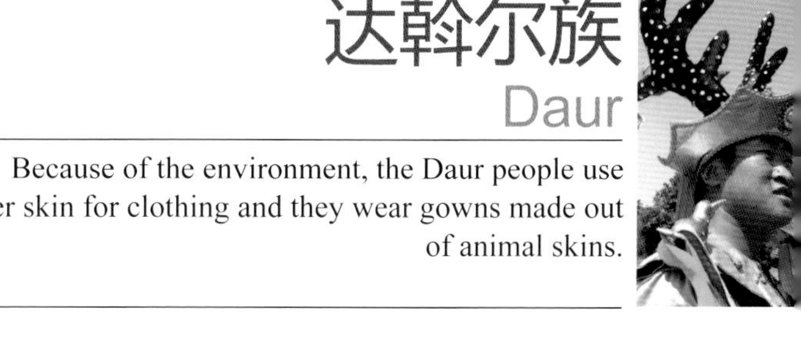

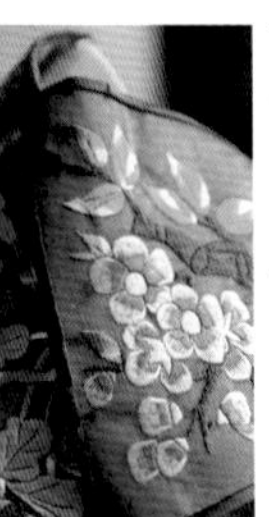

撒拉族

Salar

撒拉族传统服饰颜色鲜艳明快，上衣一般较为宽大，腰间系布。

撒拉族主要聚居在中国的青海省，人口共有13万余人；有本民族的语言，无文字。

撒拉族是信奉伊斯兰教的民族。由于生活的地区受到其他民族的影响，撒拉族服饰有两方面的特点，既具有伊斯兰教的色彩，又与回、藏、汉等民族服饰相互影响和融合。

撒拉族传统服饰中，男性多穿白衬衣搭配黑坎肩，饰腰刀，绑束腰带并身着长裤，踩布鞋；女性喜欢穿色泽鲜艳的大襟花衣，外套一件黑绿色对襟坎肩，穿长裤和绣花鞋，戴盖头。

There are about 130,000 Salar people living in Qinghai Province. Women like to wear kerchiefs on their head and black sleeveless jackets over clothes in striking colors. They are good at embroidery and often stitch flowers in five different colors onto their pillowcase, shoes and socks. Men wear white shirts covered with black vest, with belt and knife, trousers and shoes made of cotton cloth.

撒拉族刺绣
Embroidery of Salar

布朗族

Blang

布朗族妇女均挽髻于顶，挽髻处插有“三尾螺”簪，逢喜事盛会，发髻上还别有多角形银牌，髻下系有银链等装饰品。布朗族独特的织布和染色技术在我国民族织染业中独树一帜。

布朗族生活在中国的云南省，人口共有11.9万人；有自己的民族语言。

布朗族女性多穿圆领镶花边的内衣，外穿收腰宽摆及臀窄袖衣，下穿最有特点的蓝黑色印花筒裙，裙摆饰花边，挽髻缠头帕，喜戴银饰；布朗族男性多上穿无领长袖衣，下穿及脚踝大裆裤，喜欢用布包头做装饰。

布朗族独特的审美体现在：年轻妇女喜欢戴各色玻璃珠，头上佩戴鲜花，中老年妇女则爱以护腿布缠腿。

The population of this ethnic group is 119,000. The Blang people live mainly in Yunnan Province. Blang men wear collarless jackets with buttons down the front and loose black trousers. They wear turbans of black or white cloth. Women wear tight collarless jackets and tight striped black skirts. They tie hair into bun and cover it with layers of cloth.

布朗族
Blang

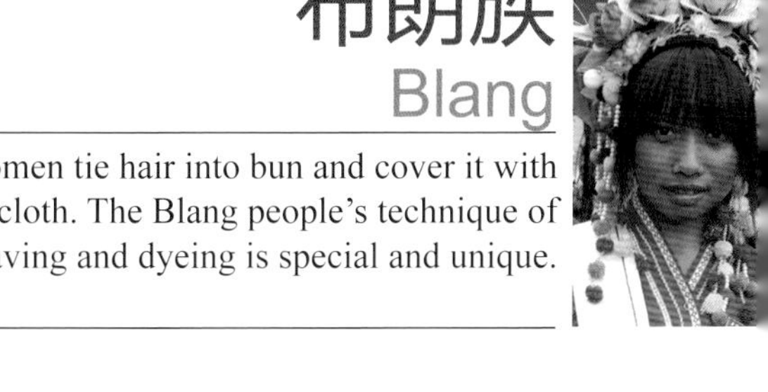

The Blang women tie hair into bun and cover it with layers of cloth. The Blang people's technique of weaving and dyeing is special and unique.

毛南族
Maonan

毛南族最具特色的服饰是由本地竹子编制的花竹帽。妇女爱穿绣鞋，有“双桥”“猫鼻”“云头”等三种款式。

毛南族生活在中国的广西壮族自治区，人口共有 10 万余人；有自己的民族语言。

毛南族传统服饰中最具特色的是花竹帽，用毛南族聚居地盛产的金竹和墨竹编制而成，是美丽、幸福的象征。

毛南族女性多穿高领上衣和紧头长裤，衣裤袖口都会绣有三道花边，腰前围黑色绣花小围裙，头梳辫，已婚戴头巾；男性上穿低领钉有五枚铜扣的上衣，下着宽筒裤，包黑头巾。

The Maonan ethnic minority,with the population of over 100,000, live in the northern part of the Guangxi Zhuang Autonomous Region. The Maonan people prefer the color of green. Women wear tight jackets with buttons on the right side, and long trousers. There are three embroidered rings on both the sleeve and leg cuffs. Men wear jackets with five brass buttons on the right side and loose trousers and black turbans.

毛南族称花竹帽为“顶卡花”，是用当地盛产的金竹和墨竹篾子编织而成的，工艺精致，花纹美观，帽形大方，结实耐用。
The bamboo hat is the specialty made by the Maonan people.

毛南族
Maonan

The Maonan women like to wear embroidered shoes.

毛南族姑娘
A Maonan girl in traditional costumes and headgear

塔吉克族
Tajik

塔吉克族主要居住在气候寒冷的帕米尔高原，服装多用皮毛、毡褐为面料。“柯而塔勒克”帽是塔吉克妇女区别于其他民族妇女的重要特征和标志。

塔吉克族主要聚居在中国的新疆维吾尔自治区，人口共有5.1万人。塔吉克族有自己的语言。

特色饰品是塔吉克族传统文化的重要组成部分。塔吉克族妇女的帽子上有一块称“柯而塔勒克”的前檐，刺绣图案五彩缤纷，妇女着盛装的时候，帽檐上还会加缀一排小银链，同时佩戴项链和各种银质胸饰。已婚妇女在发辫上缀以白纽扣，独具民族特色。

Most of the 51,000 Tajiks live in the Xinjiang Uygur Autonomous Region. Men wear collarless long jackets with belts, on top of which they add sheepskin overcoats in cold winter. They wear tall lambskin hats, and the flap can be turned down to protect ears and cheeks from wind and snow. Married women wear black aprons, and cotton-padded hats with flaps. Women usually tie a white square towel on top of their head when they go out. Both men and women wear felt stockings and long soft sheepskin boots.

塔吉克族鼓手
Tajik young men playing handdrum

巴洛托节上，阿米尔、古兰丹姆和众人跳起了欢乐的民族舞蹈
The Tajik dancing scene

帕米尔高原上的塔吉克族妇女服饰
The Tajik women in traditional costumes and headdress on the Pamir Plateau

塔吉克族
Tajik

The Tajik people use animal skins for clothing. The tall lambskin hat is the Tajik woman's symbolic sign.

鹰笛是塔吉克族最喜欢的乐器
Eagle flute is the Tajik people's favorite musical instrument.

普米族
Pumi

普米族崇尚白色服装，男性穿短衣、阔腿裤，女性则穿白色短衣套坎肩，其头饰也有鲜明的特点。

普米族生活在中国的云南省，人口共4.2万人。普米族有自己民族的语言。

普米族崇尚白色，在传统服饰中都能看到白色，现在喜用白、黑、红等多种颜色。男性穿短衣和阔腿裤，会披白色的羊皮坎肩，或穿氆氇呢子大衣；女性则穿白色短衣，再套黑褐色绣花坎肩，围腰带，有的穿镶花边金丝边短衣配羊皮坎肩，下穿白色百褶长裙，彩带束腰。在白色的服饰上，普米族男女还会佩戴银、珊瑚、玛瑙等首饰。

The most of 42,000 Pumi people live in Yunnan Province and some in neighbouring Sichuan Province. Traditionally, the Pumi people are fond of white, and like silver earrings and bracelets. Men wear linen jackets, loose trousers and sleeveless goatskin jackets. Women often wrap their heads in large handkerchiefs, winding plaited hair. They wear short jackets with a black vest, pleated skirts and a multicolored belt.

云南省兰坪县通甸乡普米族牧民的传统服饰
The Pumi herdsman in traditional clothes

普米族
Pumi

The Pumi people like white color. The headgear is peculiar.

云南罗古箐普米族的迎宾舞
The Pumi people are dancing in traditional costumes.

阿昌族

Achang

阿昌族服饰古老而独特，与高寒山区的自然生态相适应，带有明显的山地特征。

阿昌族生活在中国的云南省，人口共有3.9万人；有本民族语言。

阿昌族传统服饰中男性经常会戴耳环，出门背“筒帕”挎包和阿昌刀；未婚女性穿短衣长裤，已婚则穿短衣筒裙，戴银质项圈耳环，用贝壳装饰衣裙。男女多着蓝、白、黑色上衣，黑色长裤，包白色或藏青色头帕。

最具特色的是阿昌族男女会在包头上插饰一朵朵鲜花，芳香美观，象征阿昌族的纯洁正直。

There are about 40,000 Achang people living in Yunnan Province. Achang men wear blue, white or black jackets with buttons down the front, and loose black trousers. Achang women wear skirts and wrap their head with black or blue cloths after marriage and girls wear trousers. Achang women like to wear silver objects on festive occasions.

户撒乡阿昌族妇女服饰
Achang women in traditional costumes

可爱的云南阿昌族小女孩
The Achang kids wearing traditional costumes

云南省德宏阿昌族小伙子的装束
The Achang young men's attires, Dehong, Yunnan Province

阿昌族

Achang

The costumes and ornaments of the Achang people are ancient and peculiar. The style is very mountainous.

云南阿昌族家里至今保留着传统的织布方法。
The traditional equipment of weaving retained till today.

怒族
Nu

由于纺织技术传入怒族地区较早，因此怒族妇女用麻线纺织制作的怒毯颇负盛名。怒族善于织麻布，因而怒族男女服装多由麻布制成。

怒族生活在中国的云南省，人口共有 3.7 万人；有本民族语言。

怒族女性喜欢织麻布，所以怒族传统服饰多用麻布制成，男女都喜欢佩戴配饰。男性多穿麻布长衫，下穿长裤裹麻布绑腿，左腰佩挂砍刀，右肩背弓弩箭包；女性穿麻布长裙、上衣，外套黑红坎肩，有的穿衣裤，裤外围两块彩条麻布，用红藤缠头和腰、胸，装饰珊瑚、玛瑙、贝壳、银币等饰品。

The Nu ethnic minority, with a population of over 37,000, live mainly in Yunnan Province and some in the Tibet Autonomous Region. The Nu women are good at linen spinning so that the Nu people wear linen clothes. They like to wear coral, agate, shell and silver coin ornaments in their hair and on their chest. Women wear skirts or trousers. Men wear gowns and shorts, and carry axes, bows and arrows.

怒族老人娅普妈用自己纺织的怒毯缝制衣裙。
A Nu woman is making linen clothes.

怒族
Nu

The spinning and weaving technique had been early introduced to Nu area, so the Nu women are good at linen spinning and they mainly wear linen clothes. The linen blanket is well known.

云南省福贡地区怒族男子传统服饰
A Nu man wearing traditional costumes and ornaments

鄂温克族

Ewenki

鄂温克族所在地区气候寒冷，衣着处处离不开毛皮。其衣服肥大、宽松、斜大襟，配束长腰带。

鄂温克族主要聚居在中国的内蒙古自治区和黑龙江省，人口共有3万余人。鄂温克族有自己的民族语言。

鄂温克族的传统服饰有典型的狩猎特色。服饰原料主要是兽皮，颜色喜欢用白色，并用布或羔羊皮制作的饰品镶边，下身多穿皮裤，着皮靴，夏天单穿皮靴，冬天则里面穿皮袜。其服饰选用的动物皮毛是十分有讲究的，要分别与季节相适应，冬天选用绒毛长密的皮毛，春秋选用小皮毛，夏天则是光板皮，起到了冬暖夏凉的作用。

The total population of the Ewenki ethnic minority is about 30,000. They live in the Inner Mongolia Autonomous Region. Traditionally, the Ewenkis use animal skins to make clothing. White is their favorite color. They wear jackets, trousers and boots, all of which are made of animal skins and felt. They choose animal skins by seasons for their clothing.

鄂温克族的节日盛装

The Ewenkis people in festival costumes

鄂温克族
Ewenki

The Ewenki people choose animal skins by seasons for their clothing. The gown they wear is loose and comfortable with a slant front.

京族

Gin

京族服装以丝绸为料，质地柔软舒适，非常适合在海边穿着。京族最有特色的装饰是斗笠。

京族主要聚居在中国的广西壮族自治区，是一个濒海而居的民族，人口共有 2.8 万人。京族有自己的民族语言，无文字。

京族的传统服饰，男性一般穿着及膝窄袖袒胸上衣，宽筒长裤，腰间系带；女性上着菱形遮胸布，外穿窄袖紧身对开襟无领上衣，下穿宽筒黑褐色长裤，头梳正中平分的“砧板髻”，两边结辫黑布缠绕盘于顶，非常喜爱戴耳环。

京族葵帽也称竹笠，是用草和竹子编成，质轻、透气、防晒及防雨。它最大的特征是圆锥形造型，可大可小的圆锥横切面，能适应各种尺寸的头。竹签是京族生活中不可或缺的必需品，是京族人的标志之一。

With a population of 28,000, the Gin ethnic minority lives in compact communities primarily on three islands near the Sino-Vietnamese border in the Guangxi Zhuang Autonomous Region. Men wear long jackets reaching down to the knees and girdles. Women wear tight-fitting, collarless short blouses buttoned in the front and broad black or brown trousers. They like earrings.

京族
Gin

The Gin people choose silk and gauze as the main materials for clothing and their costumes are simple but beautiful. The spire hat is their distinctive mark and they usually wear it to avoid the strong sunshine and rain.

基诺族
Jino

基诺族喜欢穿自织的带有蓝、红、黑色彩条的土布衣服。日月花饰是基诺族成年男子的背部装饰。基诺族妇女戴白底彩色纹的披风式三角尖顶帽。

基诺族生活在中国的云南省，人口共有2.3万人；有自己的民族语言。

基诺族女性善于纺织，最具特色的是系在腰间自己纺织出来的砍刀布。用砍刀布做成的披风式尖顶帽是基诺族最具特色的服饰。

基诺族喜欢赤脚，穿戴耳饰，男性穿黑白格子上衣，衣服前绣红蓝花条，后背绣太阳花纹，下穿宽大裤子；女性穿鸡心式绣花胸兜，配一件红蓝白黄花格上衣，下穿红布镶边短裙，裹蓝黑色绑腿。

With a population of 23,000, the Jino people live in Yunnan Province. The Jino men usually wear collarless white jackets and white or blue trousers. Women wear multicolored and embroidered collarless gowns and short black skirts rimmed in red and opened at the front.They also wear their hair in a coil just above the forehead, and sling across shoulders sharp-pointed flax hats.

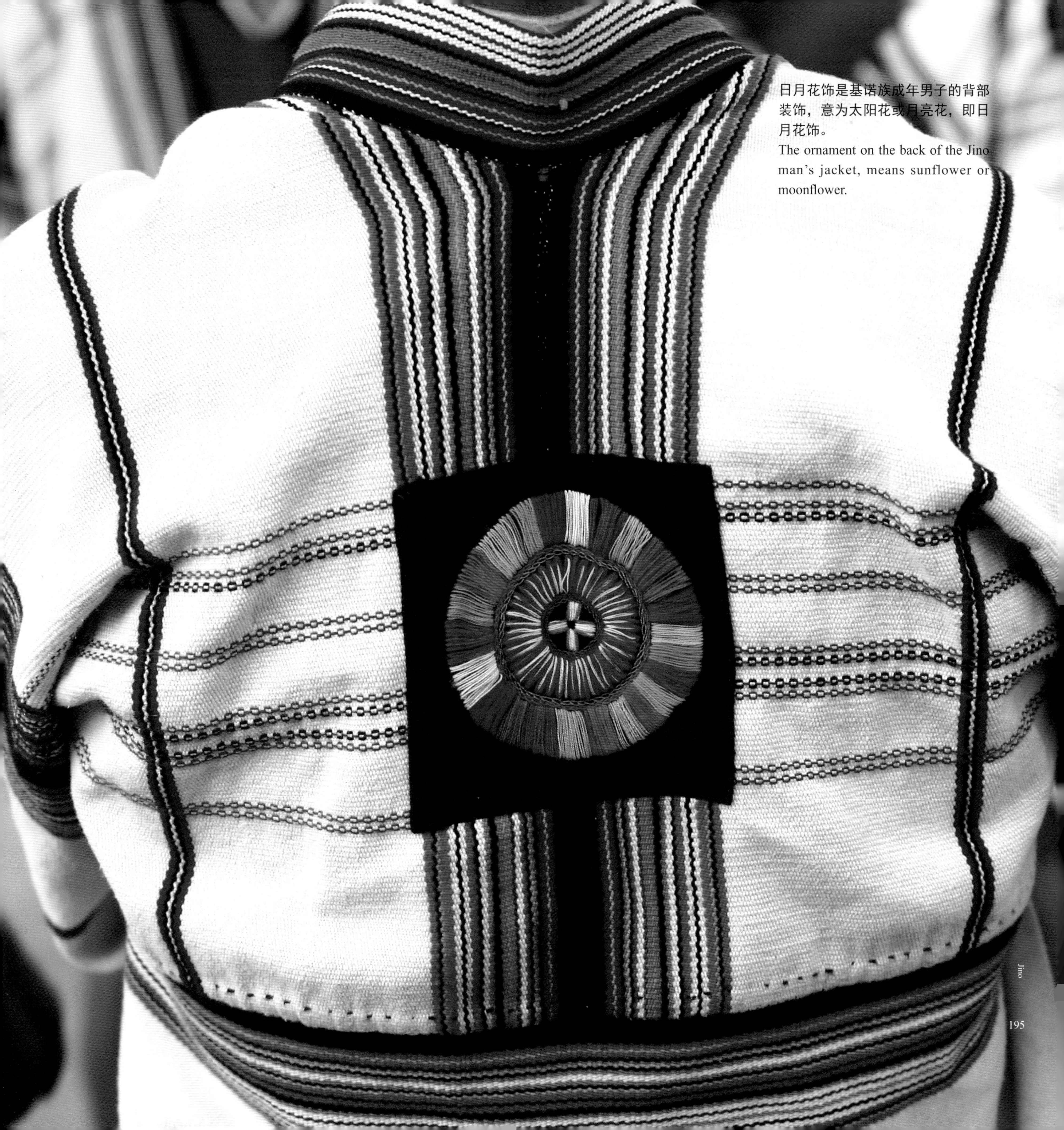

日月花饰是基诺族成年男子的背部装饰，意为太阳花或月亮花，即日月花饰。

The ornament on the back of the Jino man's jacket, means sunflower or moonflower.

基诺族妇女的尖顶帽别具一格。
The peaked hat of the Jino woman is very remarkable.

基诺族
Jino

The Jino people like to wear costumes of their own made cotton cloth with black, blue and red stripes. The patterns of the sun and the moon on the back of the man's jacket and the woman's peaked hat are distinctive.

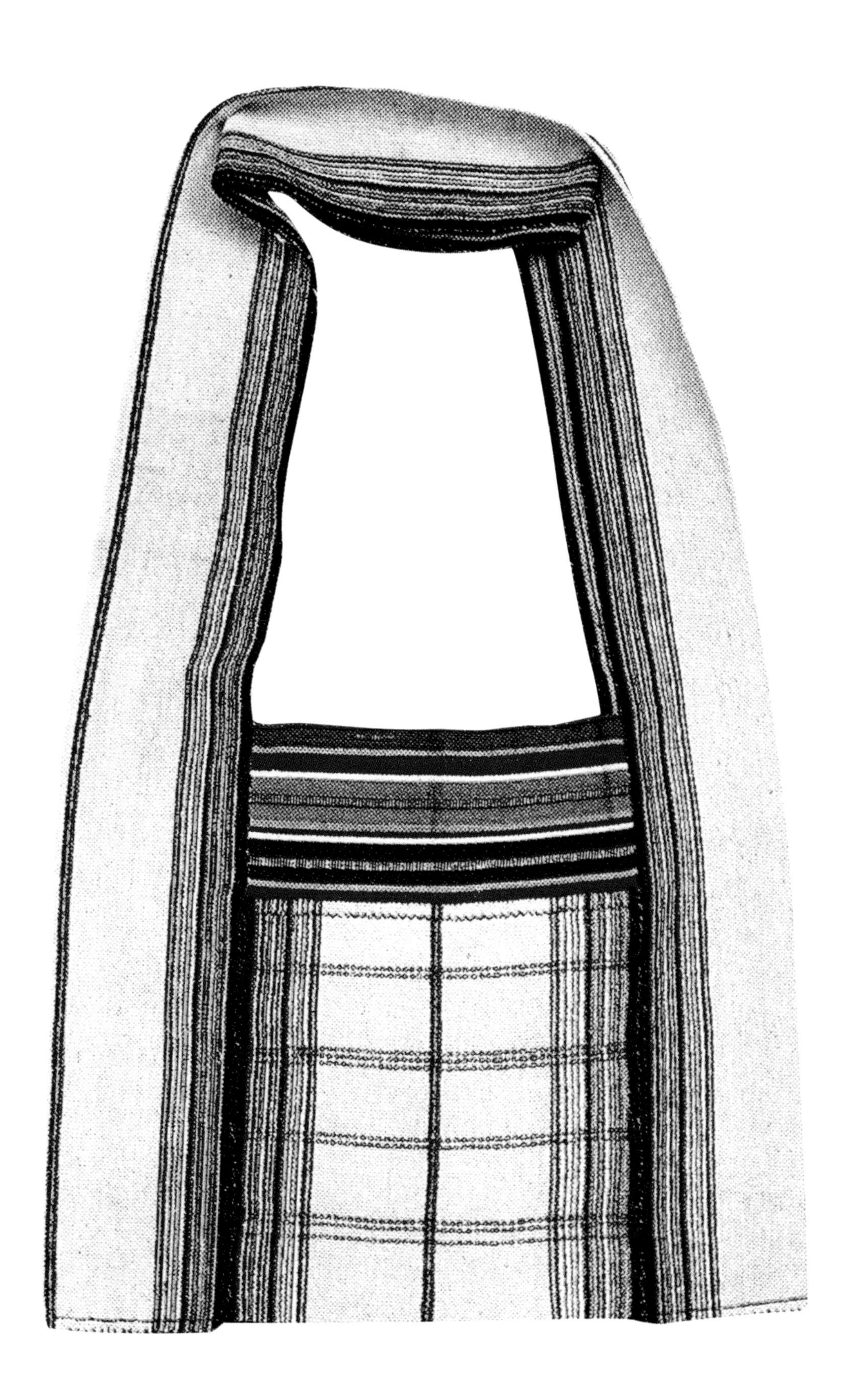

基诺族自织条纹布挎包
A shoulder bag made with striped cloths.

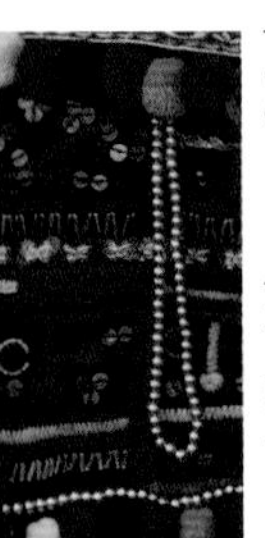

德昂族

De'ang

在德昂族的服饰中，最引人注目的是妇女身上的腰箍，妇女们以佩戴腰箍多为荣。德昂族妇女彩条水波横纹筒裙的图腾纹饰反映了其先民对龙和大鸟的崇拜。

德昂族主要聚居在中国的云南省，是中国西南边疆最古老的世居民族之一，人口共有 2 万余人。德昂族有自己民族的语言，无文字。

德昂族女性穿蓝黑上衣，衣边镶红布条，用大方银牌为纽扣，成年妇女一般剃发，并用黑布包头两端后垂。传统服饰中最有特点的是女性腰间缠的各种花纹藤篾圈，用藤篾、银丝编成，刻上各种花纹图案，包上银皮或铝皮，这是德昂族从先民延续而来的服饰传统。德昂族男性多穿蓝黑上衣、宽短裤，用黑白布包头，两端饰各种绒球，带大耳坠和银项圈。

The number of De'ang people totals about 20,000, and they live in Yunnan Province. De'ang men usually wrap their heads with black or white scarves.They like to wear silver ornaments on their ears, necks and hands. They wear short black or blue jackets and loose trousers. Women wear black or blue blouses with red stripes of cloth on edges and four or five big square silver buttons. They wear bamboo girdles with delicate ornamental designs.

德昂族妇女五彩绒球饰

Colorful pompon ornaments on De'ang women's costumes

德昂族

De’ang

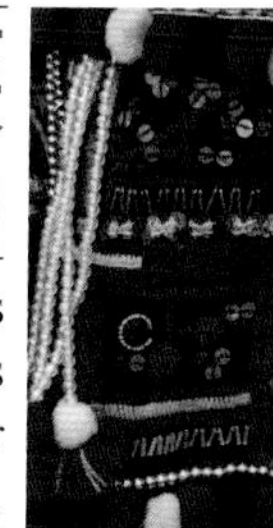

The most distinctive part of De’ang people’s costumes is the waistband for women. The patterns on the band reflect De’ang people’s respect to their ancestors.

黑德昂支系妇女

De’ang women in their traditional costumes

保安族
Bonan

保安族服饰最具民族特色的是“保安刀”佩饰，已经有 100 多年的历史，不仅是生活用具，也是别致的装饰品和馈亲赠友的上乘礼品。

保安族生活在中国的甘肃省，人口共有 2 万余人。保安族有自己的语言，无文字。

保安族传统服饰因地而变，早期均以穿长袍为主。这是一个喜欢佩带腰刀的民族，腰间必挂保安族最具特色的配饰——腰刀。每户保安族人家都有铁匠，每位铁匠都有着自己独特的腰刀设计图案，有的图案蕴含着美好的传说，或记载着一个悲壮的故事。这既是男女爱恋的信物，又是馈赠乡亲的礼品。保安族女性传统服饰戴头盖，穿大襟上衣，外穿艳色镶边坎肩。

This ethnic group in Gansu Province, with a population of about 20,000, is one of China's smallest minorities. Traditionally, Bonan people like to wear gowns and hats made of sheep skins. Bonan men like wearing knives, which are famous all over China for their beauty and hardness. The Bonan knife is used not only as a gift, but also as a token of love.

保安族

Bonan

The Bonan men are proud of their knives which are seen as one of the three famous knives made by China's minority ethnic groups. The knife is also a nice ornament and gift.

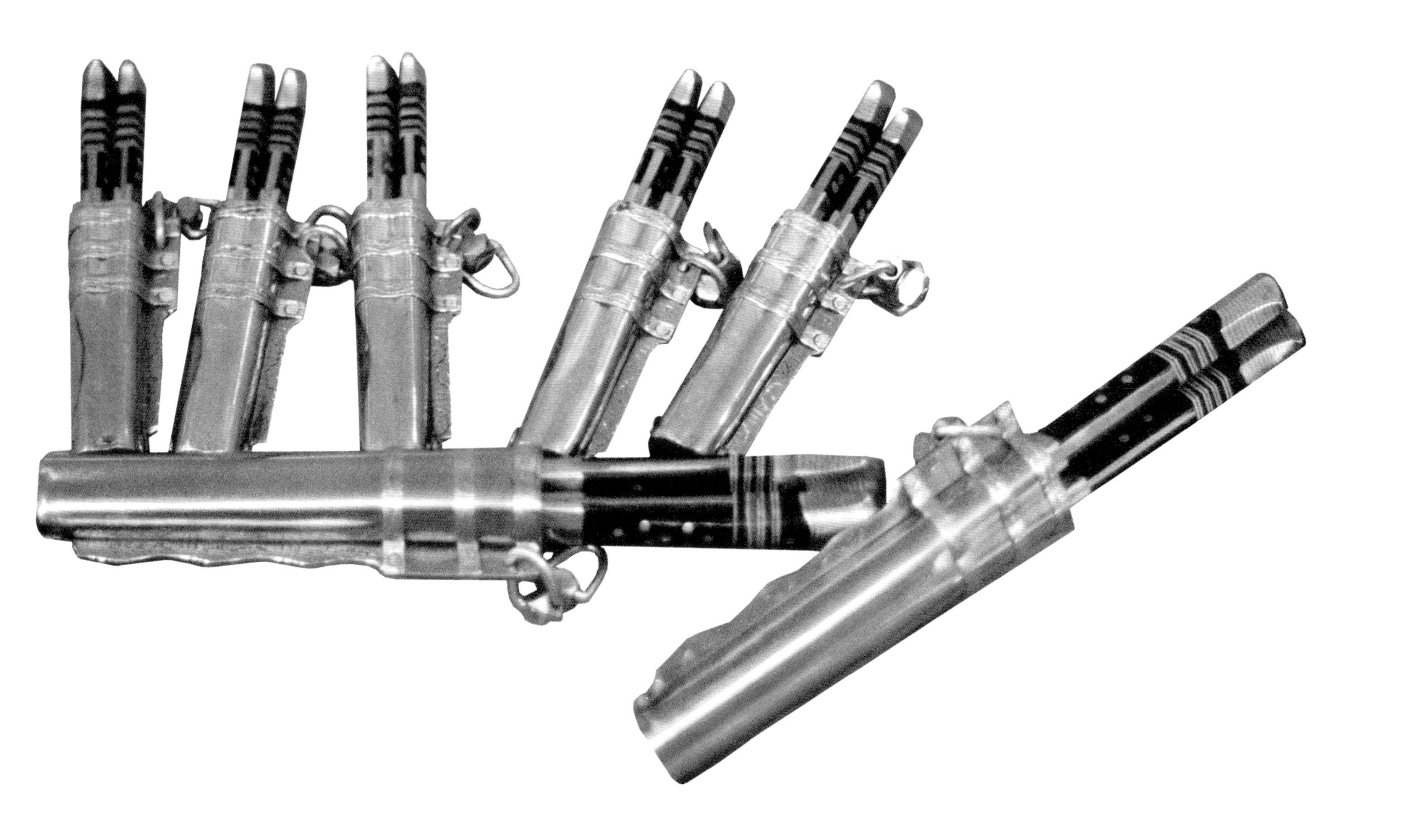

保安腰刀是中国少数民族三大名刀之一
Bonan knives

俄罗斯族
Russian

俄罗斯族的服饰丰富多彩，人们在不同季节里，选择不同颜色、不同款式的衣着。俄罗斯族妇女的头饰颇具特色，年轻姑娘与已婚妇女的头饰有严格区别。

俄罗斯族主要聚居在中国的新疆维吾尔自治区，人口共有1.5万人。俄罗斯族有自己的民族语言，并使用俄文。

俄罗斯族的传统服饰，男性上穿斜领麻布衬衫，外披呢子，系腰带上衣，下着细腿长裤，穿长筒皮靴；女性上着领口带褶粗麻衬衫，外穿“萨拉凡”无袖长袍或家织毛裙。

There are about 15,000 ethnic Russians living in the Xinjiang Uygur Autonomous Region. Their clothing is almost identical to that of the Russians in Russia.

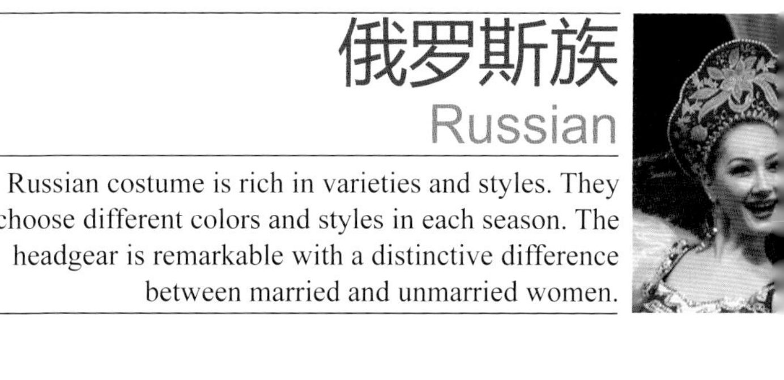

俄罗斯族
Russian

Russian costume is rich in varieties and styles. They choose different colors and styles in each season. The headgear is remarkable with a distinctive difference between married and unmarried women.

俄罗斯族妇女服饰的衣领、袖口和前胸等部位缀精美细密的刺绣几何图案或花草图案，色彩鲜艳，对比强烈。
Embroideries in patterns of geometric shape and flowers are used for decoration on costumes.

裕固族
Yugur

裕固族游牧于戈壁，逐水草而居，世代以畜牧业为主，服饰用料取之于畜牧业本身，衣着式样蕴涵游牧民族特色的服饰文化。

裕固族主要聚居在中国的甘肃省，人口共有1.4万人。各地裕固族使用三种不同的语言，使用汉字。

裕固族传统服饰有着游牧民族的特征：高领、白皮袄、长筒皮靴，耐寒防沙，方便放牧时穿着。

裕固族世代以畜牧业为生。褐子是裕固人用来缝制衣物、褡裢、帐篷的手工粗布，有良好的防水、避风、隔潮、耐晒、保温作用。织褐子的原材料是用羊毛、驼毛、牦牛毛等手工捻成的不同用途的毛线。用褐子制作的各种生产、生活用品，是裕固人最常用、最传统的手工艺品。

The Yugur ethnic minority, with a population of 14,000, live in a Yugur autonomous county in Gansu Province. The Yugur people have their own peculiar way of dressing. A typical well-dressed man wears a felt hat, a high-collared long gown buttoned on the left, a red-blue waist band and high boots.The women usually wear a trumpet-shaped white felt hat with two black lines in the front, topped by red tassels.

裕固族女子服饰
Yugur women in traditional costumes and headgear

裕固族
Yugur

The Yugur costume is typical of nomadic people, sheep skins and high-collared long gowns with waist bands and high boots.

裕固族妇女在捻羊毛
Yugur women twisting wool

乌孜别克族
Uzbek

乌孜别克族男女都戴各式各样的小花帽。妇女们戴花帽时，常在小帽外再罩上薄如蝉翼的花色纱巾，别有一番风韵。

乌孜别克族主要聚居在中国的新疆维吾尔自治区，人口共有1万余人。他们有自己的民族语言，使用阿拉伯字母文字。

早在14世纪的时候，乌孜别克族在丝绸之路的影响下，经营丝绸、茶叶、瓷器、皮张、大黄和各种土特产。乌孜别克族传统服饰中，夏天通常着丝绸所制的衣服，男性着套头短袖衬衫，领口、袖口、开襟用红绿蓝丝线绣花边，腰间系绸缎或棉布的三角绣花腰带；女性喜爱穿多褶连衣裙，戴乌孜别克花帽。

The Uzbek ethnic minority, with a population of 14,800, are scattered over wide areas of the Xinjiang Uygur Autonomous Region. In summer time, both men and women wear clothes made of silk, pullover shirts and skirts. They wear warm sheep skin clothes for winter.

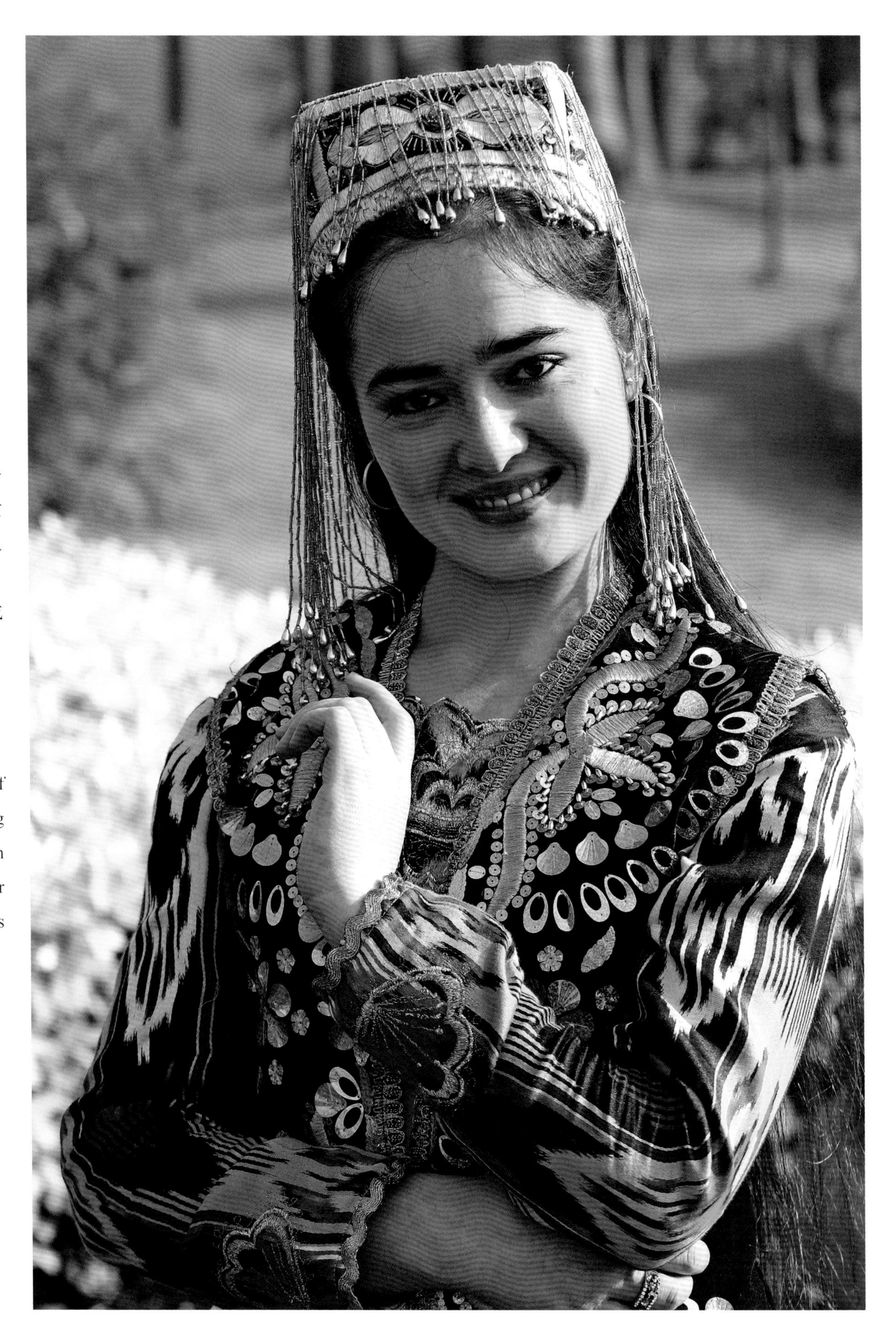

乌孜别克族
Uzbek

The Uzbek people are fond of wearing caps. Colorful caps in various styles are peculiar.

门巴族
Menba

门巴族由于长期与藏族相处，在吸收较多藏族文化的同时，也形成了自己独特的服饰文化和传统习俗。

门巴族生活在中国的西藏自治区，人口共有1万余人；有自己的民族语言，通用藏文。

门巴族喜欢穿藏式氆氇长袍，穿牛皮靴子。传统服饰中最具特色的是帽子，无论男女都喜欢戴“拔尔甲”小帽。帽子用蓝黑色与红色氆氇做成，帽檐是黄褐色绒布边，并且要留一缺口，戴帽子时，男性缺口在右眼上方，女性缺口往后。女性用红绿黄线梳辫盘头，佩戴红珊瑚、绿松石、银质首饰，男性蓄长发佩松耳石耳饰。

Menba People, numbering about 10,000, are scattered in the southern part of the Tibet Autonomous Region. The Menba people prefer to wear robes with aprons and black-yak-hair hats or caps. They wear soft-soled leather boots. They usually wear earrings, rings and bracelets. Women wear long striped skirts and various kinds of jewelry.

西藏山南地区门巴族妇女服饰

The costumes of Menba women in Shannan Prefecture

门巴族
Menba

The Menba people live in the same area with Tibetans, so their costumes are quite similar but with some peculiar signs in decoration which reflect their culture and tradition.

门巴族男子传统服饰
The traditional costumes of Menba men

鄂伦春族
Oroqen

鄂伦春族的服饰充分显示了狩猎民族的特色。鄂伦春妇女加工的狍皮结实、柔软、轻便，为了适应寒冷气候和狩猎生活所创制的狍皮衣和狍皮帽独具匠心，别具特色。

鄂伦春族主要聚居在中国的内蒙古自治区和黑龙江省，祖祖辈辈生活在大兴安岭的原始森林中，是一个狩猎的民族，共有8000余人，使用自己的民族语言和汉语。

受居住环境影响，森林里的动物狍子皮成为鄂伦春族人传统的主要衣料，包括帽子、衣、裤、手套、靴子等。一身狍皮所制的猎服，既能在冬天抗寒、夏季防潮，耐磨经扯，方便温暖，又是最佳狩猎武装。

There are about 8,000 Oroqen people dwelling in the forests of the Greater Khingan Mountains in Northeast China that abound in deer and other beasts, which the Oroqens hunt with shot-guns and dogs. The deer skin is the main fabric for garments they wear to keep warm in winter, damp-proof in summer and durable in hunting. The pelt prepared by women are also soft, fluffy and light for making hats, gloves, socks and blankets as well as tents.

内蒙古呼伦贝尔鄂伦春族妇女手工缝制狍子皮
The Oroqen women are sewing roe deer skin in Hulun Buir, Inner Mongolia.

鄂伦春族
Oroqen

The Oroqen people live in forests and hunting is their main production for living. They like costumes in deer skin which are suitable for the weather and their living.

鄂伦春族的围猎
The Oroqen hunters

独龙族

Derung

独龙族的装束相对简单，白天作披肩、晚上当被盖的独龙毯是他们最重要的传统民族服饰。

独龙族生活在中国的云南省，人口共有6000余人；有自己的民族语言。

独龙族传统服饰中最具特色的地方是男女都斜披自织麻布衣独龙毯，白天为衣，夜晚为被，既实用也起到装饰作用。此外，独龙族多下穿麻布短裤，男女都赤脚蓄发，喜爱装饰，常披挂彩色珠串、胸链、耳环，男性腰系弩弓，女性出门背篾篓。背篾篓和独龙毯都体现了独龙族传统服饰中对于实用性的重视。

The Derung people, numbering about 6,900, live mainly in Dulong ıiver valley of the Gongshan Derung and Nu Autonomous County in northwestern Yunnan Province. Derung people are good at making their special clothing, the Derung blanket. People like to wear a very beautifully striped linen robe which could be used to cover as a blanket while sleeping. Both men and women go in bare feet. They like to wear earrings, bracelets and ornaments.

独龙族 Derung

The Derung people wear simply and the blanket is important to them. They use blankets to wrap around in the daytime and cover with them at night.

赫哲族

Hezhe

渔猎生活给赫哲族服饰打上特别印记，他们早年服装的主要原料是鱼皮。赫哲族的鱼皮服饰文化世所罕见。

赫哲族世代生活在中国的黑龙江省，目前人口共有5000余人。他们使用自己的民族语言和汉语。

赫哲族聚居的三江流域水产资源十分丰富，孕育了他们独一无二的鱼皮服饰文化。他们用鱼皮作为服装的面料，鱼筋作为缝制衣服的线，用鱼骨磨制作为扣子，整件服饰材质均取之于鱼，因而服饰御寒抗湿、防水耐磨。这是赫哲族对人类服饰文化做出的特殊贡献。

Hezhe people,who make up China's smallest ethnic group (population of over 5,000), live in North China, around the three major river valleys. The Hezhe people live by hunting and fishing so that the garments they wear are mostly made of fish or deer skins, among which the fish-skin dresses are the most unique.The process of making fish-skin clothing is very interesting and considered as their special contribution to the collection of China's colorful ethnic costumes.

赫哲族
Hezhe

The Hezhe people live on fishing. They make use of fish skins. Their way of wearing fish-skin clothes is rarely seen.

江中鱼，人之衣——赫哲族鱼皮服饰
Costumes made of fish skins

台湾少数民族
Ethnic Minorities in Taiwan

台湾少数民族传统服饰色彩鲜艳，以红、黄、黑三种颜色为主。除服装外，还有许多饰物，如冠饰、臂饰、脚饰等，以鲜花制成花环，在盛装舞蹈时，直接戴在头上，非常漂亮。

台湾少数民族包含泰雅、赛夏、布农、邹族、鲁凯、排湾、卑南、阿美等 16 个族群，2014 年的人口统计为 539435 人，主要聚居在中国台湾，在大陆被称为高山族。

台湾少数民族男女都喜欢佩戴精美的饰物，其服饰色彩斑斓，做工讲究，充分展示了他们艺术创造的才能。根据各族群的分布区域不同，服饰也有不同风格。他们的主要衣料由麻布、皮革和棉布制成。擅长织布的泰雅人的织技高超，贝珠衣是他们十分珍贵的服饰——一件讲究的贝珠衣上的装饰最多有十几万颗贝珠。

The ethnic minorities in Taiwan comprise 16 tribes and the total population is 539,435. They are known as Gaoshan ethnic group.

The majority of them live in mountainous areas and the flat valleys running along the east coast of Taiwan island. There are about 1,500 people residing in Beijing, Shanghai, Hubei and Fujian. Their clothing is colorfully designed and well made. Men usually wear sleeveless jackets, top and belts, and some wear deerskin waistcoats, waist band and black-cloth skirts. Women wear long skirts with short jackets or short jackets with long skirts. They like to wear ornaments and decorate the whole body.

台湾少数民族　赛夏青年男女服饰背面带有臀铃
Saisiyat youth in Taiwan wearing bells at the back as the costume decoration

台湾少数民族　阿美人
The people of Amis in Taiwan

台湾少数民族　阿美人和鲁凯人服饰
Amis people and Rukai people in Taiwan

台湾少数民族　鲁凯人服饰
The people of Rukai in Taiwan

台湾少数民族　排湾人服饰
The people of Paiwan in Taiwan

台湾少数民族

Ethnic Minorities in Taiwan

The people of ethnic minorities in Taiwan are fond of colorful costumes. They prefer red, yellow and black. They like to wear ornaments and decorations. They wear garlands when dancing. The scene is gorgeous.

珞巴族

Lhoba

珞巴族充分利用野生植物纤维和兽皮为原料，显示出山林狩猎生活的特色。最富特色的装饰是腰饰和耳鼓。

珞巴族生活在中国的西藏自治区，人口共有3000余人，是中国人口最少的民族之一。珞巴族有自己的民族语言，通用藏文。

珞巴族男女均赤脚，留长发，额前头发剪齐于眉际以上，其余披散肩后，佩戴铜银、海贝等饰品。为方便狩猎和农事，男性穿藏式氆氇长袍，背披牛皮，头戴熊皮圆盔；女性穿无领上衣，外披小牛皮，下着过膝紧身筒裙。

The 3,600 people of the Lhoba ethnic minority have their homes mainly in southeastern Tibet. Lhoba men wear sleeveless, buttonless knee-length wool jackets. They wear helmet-like hats either made of bear skins or woven from bamboo strips. Women have narrow-sleeved blouses and skirts of sheep's wool.They wear a variety of waist ornaments, such as shells, silver coins, iron chains and belts.

珞巴族
Lhoba

Because the Lhoba people are hunting for a living, they make use of animal skins as well as fibers found in mountains for their clothing. Their costume decorations are remarkable.

珞巴族男子世代以狩猎为生
Lhoba men live off hunting for generations.

塔塔尔族

Tatar

刺绣是塔塔尔族妇女最擅长的技艺之一，各种服饰上都绣有多姿多彩的花纹和图案。

塔塔尔族主要聚居在中国的新疆维吾尔自治区，人口共有 3000 余人；有本民族语言。

塔塔尔族的传统服饰中男女都穿皮制鞋靴。女性喜欢穿“塔裙”，穿宽大的紧腿裤和宽大的下边带皱边的连衫长裙子，颜色多为白、黄、酱色，外披西服上衣或深色坎肩，并喜欢佩戴金耳环、银耳环、手镯、戒指、项链等。

There are about 3,000 Tatars in China, most of whom live in the Xinjiang Uygur Autonomous Region. Tatar people are fond of black and white in color. Men usually wear embroidered white shirts and short black vests or long gowns, embroidered black and white hats or black fur hats in winter. Women wear small hats inlaid with pearls and long dresses with pleats. They like to wear jewelry and put silver coins on their clothing.

塔塔尔族
Tatar

The Tatar women are good at embroidery, and decorate their costumes with various embroidered patterns.

后记

《丝路霓裳》（中国卷）是中国丝绸之路沿线各民族绚丽多彩服饰的集中呈现。本书以图文并茂、汉英对照的形式，集中展现中国各民族服饰文化绚丽多彩之美；立足于当代视角，重在体现中华服饰文化的传承以及与国外服饰文化的交融和创新发展。本书旨在引领海外读者用更多彩的目光来认识中国各民族的文化，领略中华民族服饰的精髓和独特文化魅力。美不胜收的民族服饰，折射出延续数千年的民族服饰文化所体现的各族人民对真、善、美的不懈追求，各民族文化的交融互鉴，引领读者跨越时空，触摸中华服饰文化的脉络与血肉，给人以最直观的视觉美感体验。

本书的出版，得到了许多业内专家的大力支持，也得到了国家民委、外交部、文化与旅游部、中国人类学民族学研究会等单位领导的关心和指导，在此深表感谢。安徽人民出版社的领导对本书的出版非常关心和支持，协调编辑、出版、发行等部门积极配合，使本书得以高效率出版。在此，还要感谢《民族画报》社央金女士对图片的统筹，感谢张体斌老师精准地将书稿及时翻译成英文，感谢苗族艺术家徐人杰先生对装帧设计的倾心投入，感谢保利艺术博物馆赵晓雪女士对全书文字的统筹，感谢中国人类学民族学研究会博物馆专业委员会的吕保利和胡良友等专家提供的帮助。

本书中有个别图片的作者不详，由于编纂时间较紧，编委会多方联系未果，在此向您致歉，并希望作者与编委会主动联系，我们将按国家标准支付图片稿酬。

《丝路霓裳》（中国卷）编委会

2019 年 4 月

Postscript

This album offers an exhibition of colorful and diverse Chinese folk arts in clothing culture of China's 56 ethnic groups. With each costume reflecting the spirit of the pursuit of happiness, opening and inclusiveness, we wish that our readers could feel the distinctive history of Chinese clothing culture.

When we started our compilation of this album, we aimed to make it a book to show colorful ethnic costumes of China and to display the history of the development of China's clothing. This album can also be used as a reference book for garment designing as well as a good choice for a gift.

Many hands and cooperation has made this album possible. Acknowledgments are extended to those help from the Government ministries and departments, to Anhui People's Publishing House, Time Publishing and Media Co.,Ltd. Without the collaboration, this album could not be published in a way of high efficiency and exquisite quality. High appreciation is also given to Ms. Yang Jin of the Nationality Pictorial for her wonderful work in selection and coordination regarding the beautiful photos, to Ms. Zhao Xiaoxue of the Poly Art Museum for her wonderful work on the literature of this book, to Mr. Zhang Tibin who translated the literal text. Mr. Xu Renjie who is an artist of Miao nationality deserves the special acknowledgement and high appreciation for his dedication to the design of this album. Thank Dr. Hu Liangyou and the group of scholars of anthropology from the Committee of Museology of the CUAES.

Editorial Committee

April, 2019

《丝路霓裳》（中国卷）编委会

Editorial Committee

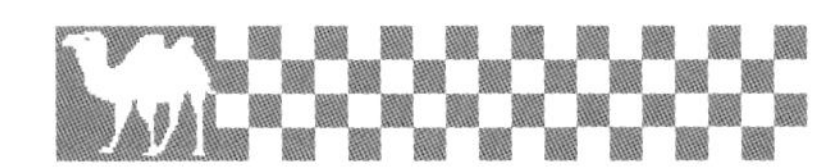